Motor Vehicle Technology
Associated Studies 2

by the same author

Auto Mate

Fundamentals of Motor Vehicle Technology
(with F.W. Pittuck)

Motor Vehicle Technology: Associated Studies 1

Motor Vehicle Basic Principles

MOTOR VEHICLE TECHNOLOGY

Associated Studies 2

V. A. W. Hillier T.Eng. (CEI), FIMI, AMIRTE

*Senior Lecturer in Automobile Engineering,
Croydon College*

Hutchinson
London Melbourne Sydney Auckland Johannesburg

Hutchinson Education

An imprint of Century Hutchinson Limited

62-65 Chandos Place, London WC2N 4NW

Century Hutchinson Group (Australia) Pty Ltd
PO Box 496, 16-22 Church Street, Hawthorn, Melbourne,
Victoria 3122

Century Hutchinson Group (NZ) Ltd
32-34 View Road, PO Box 40-086, Glenfield, Auckland 10

Century Hutchinson (SA) (Pty) Ltd
PO Box 337, Bergvlei 2012, South Africa

First published 1975
Reprinted 1979, 1981, 1982, 1984, 1986

© V.A.W. Hillier 1975

Set in IBM Univers

Printed and bound in Great Britain by
Anchor Brendon Limited, Tiptree, Essex.

ISBN 0 09 120011 3

CONTENTS

PREFACE

Volume 1 of this title covered the requirements of the Part 1 examinations set by the City and Guilds of London Institute and Regional Examining Bodies for a newcomer to the motor vehicle repairing industry.

The material for this volume is based on the syllabus entitled *Motor Vehicle Craft Studies Part* II and covers the Associated Studies work recommended for an apprentice motor vehicle mechanic who is engaged on the repair of light or heavy vehicles. Although the book follows the syllabus published by the City and Guilds of London Institute, it is intended that the content will also be helpful to a reader who is following a similar course of study.

Each subject has been analysed to ascertain the treatment required to illustrate or reinforce the basic principles of motor vehicle technology. In general, no academic topic has been included unless its relevance to motor vehicle work has been proved. With this fact in mind it is hoped that the reader will be excused many hours of study on subjects which are not essential to his chosen career.

The SI International System of Units is now adopted by all examination authorities, so most of the material in this book is presented in SI form. Some examples have been included which use the 'bar' as a unit of pressure: this work is to cover the situation where a manufacturer is using this non-preferred unit. The relationship between Imperial and SI units has already been covered in Volume 1 so this has not been repeated. In cases outside this book, where a value has been stated in Imperial units then the conversion factors given at the back of the book will enable it to be changed to the SI value.

Many of the topics complement the material in the book *Fundamentals of Motor Vehicle Technology.* When a cross-reference is shown, the title is abbreviated to *F of M.V.T.* Subject material included in the Associated Studies section of the syllabus which has been covered adequately in the Technology book has not been repeated in this volume.

A feature of the book is the widespread use which is made of multiple choice objective items to consolidate each section of work. Over 300 items, similar to those used in external examinations, have been included to enable the prospective candidate to acquire experience in the technique of answering this type of question.

Some examination papers still contain the traditional type of question, so these are also used at the end of those chapters which have a content suitable for testing in this manner. All examples and questions apply to motor vehicles and have been selected to demonstrate a specific principle. The use of simple values in all cases shoul avoid complication and the need for mathematical aids such as logarithms.

V. A. W. Hillier
1974

CALCULATIONS

.1 SI units

There are seven basic SI units, as follows

kilogram (kg) — mass
second (s) — time
ampere (A) — electrical current
kelvin (K) — temperature
candela (cd) — luminous (light) intensity
mole (mol) — amount of substance

From these basic units others can be derived; these are units which are products or quotients of two or more base units, e.g. metres per second. Some derived units are named after famous scientists; for example

joule (J) — energy
newton (N) — force
watt (W) — power

Prefixes are used in front of the basic unit to avoid writing in full a number which consists of many digits; e.g. it is easier to write 52 km than 52 000 m. These prefixes are shown in Table 1.

Table 1. Prefixes for SI units

Prefix	Symbol	Standard Value	Value (written in full)
tera	T	10^{12}	1 000 000 000 000
giga	G	10^9	1 000 000 000
mega	M	10^6	1 000 000
kilo	k	10^3	1 000
hecto	h	10^2	100
deca	da	10	10
deci	d	10^{-1}	0·1
centi	c	10^{-2}	0·01
milli	m	10^{-3}	0·001
micro	μ	10^{-6}	0·000 001
nano	n	10^{-9}	0·000 000 001
pico	p	10^{-12}	0·000 000 000 001
femto	f	10^{-15}	
atto	a	10^{-18}	

Expressed as a 'power of 10' the value is much shorter; it would need a wider page to write in full the value of 'femto' and 'atto'. The number that indicates the power to which 10 is raised is called an 'index' (plural, indices).

The International Standards Organization (ISO) recommend that those units whose prefixes denote 10 raised to the power three, or multiple of three, should be used whenever possible to reduce the risk of errors. This does not mean that the prefixes hecto, deca, deci and centi must not be used: in practice you will find many instances where these prefixes can be used to advantage to simplify calculations. In such cases special care must be used when converting the values back to the preferred form.

Examples

Change

1. 6·5 metres to millimetres

$1 \text{ m} = 10^3 \text{ mm or } 1000 \text{ mm}$

Answer will be larger so multiply 6·5 by 10^3

6·5 m = 6500 mm   (decimal point moved three places)

2. 0·524 megametres to metres

$1 \text{ Mm} = 10^6 \text{ m or } 1\,000\,000 \text{ m}$

Answer will be larger so multiply 0·524 by 10^6

0·524 Mm = 524 000 m   (decimal point moved six places)

3. 7246 grams to kilograms

$1 \text{ kg} = 10^3 \text{ g or } 1000 \text{ g}$

Answer will be smaller so divide 7246 by 10^3

7246 g = 7·246 kg   (decimal point moved three places)

4. 56·2 milliampere to amperes

$1 \text{ A} = 10^3 \text{ mA or } 1000 \text{ mA}$

Answer will be smaller, so divide 56·2 by 10^3

56·2 mA = 0·0562 A   (decimal point moved three places)

In these examples it will be seen that the index shows the 'number of places' that the decimal point is to be moved; the direction of movement can be judged by the 'size' of the required answer.

Exercises 1.1

Change:

1.	35 mm	to	m	11.	57 m	to	dm
2.	22 kg	to	g	12.	342 mm	to	dm
3.	482 g	to	kg	13.	87·25 cm	to	m
4.	47 J	to	kJ	14.	962·5 dm	to	m
5.	47 000 W	to	MW	15.	123·5 dm	to	km
6.	0·05 km	to	mm	16.	0·512 Mm	to	km
7.	0·007 MN	to	kN	17.	32·6 hm	to	dam
8.	0·02 A	to	mA	18.	0·0061 Gm	to	km
9.	0·004 dam	to	m	19.	3628 μm	to	mm
10.	0·36 cg	to	mg	20.	7956·2 km	to	Mm

21.	64 m	to	mm	31.	48 dm	to	m
22.	92 g	to	kg	32.	874 dm	to	mm
23.	325 kg	to	g	33.	32·18 m	to	cm
24.	36 mJ	to	J	34.	825·8 m	to	dm
25.	54·3 MW	to	W	35.	645·2 km	to	dm
26.	0·08 mm	to	km	36.	0·768 km	to	Mm
27.	0·006 kN	to	MN	37.	91·3 dam	to	hm
28.	0·05 mA	to	A	38.	0·0084 km	to	Gm
29.	0·008 m	to	dam	39.	6452 mm	to	μm
30.	0·39 mg	to	cg	40.	8294·6 Mm	to	km

.2 Mensuration

Mensuration involves the rules for finding lengths, areas and volumes. Some of these rules have been covered in previous work; these included:

Area

Rectangle: $l \times b$

Triangle: $\frac{1}{2} b \times h$

Circle: πr^2 or $\frac{\pi d^2}{4}$

Volume

Rectangular solid: $l \times b \times h$

Cylinder: $\pi r^2 l$ or $\frac{\pi d^2 l}{4}$

Although most practical applications of mensuration involve the rectangle, triangle and circle, occasions arise where other shapes are used. The following shows some typical examples:

Parallelogram

A four-sided figure whose opposite sides are parallel (Figure 1.2.1).

This figure can be converted into a rectangle by cutting at AE and moving the triangle AEC to the other end of the figure.
Area = length x perpendicular height or AB x AE.

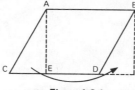

Figure 1.2.1

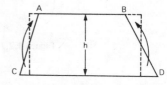

Figure 1.2.2

Trapezium

A four-sided figure having two parallel sides (Figure 1.2.2).

A rectangle is obtained by rearranging the figure as shown.

Average length of sides AB and CD = $\frac{AB + CD}{2}$

Area = average length of parallel sides x perpendicular height or

$$\frac{AB + CD}{2} \times h$$

Hexagon

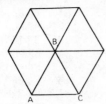

Figure 1.2.3

A six-sided figure—a regular hexagon has six sides of equal length (Figure 1.2.3).

The area of a hexagon may be determined by dividing the figure into a number of triangles.

Area of regular hexagon = 6 x area of triangle ABC.

Annulus

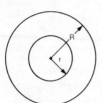

Figure 1.2.4

The figure formed between two concentric circles (Figure 1.2.4).

$$\text{Area of large circle} = \pi R^2$$
$$\text{Area of small circle} = \pi r^2$$
$$\text{Area of annulus} = \pi R^2 - \pi r^2$$
$$= \pi (R^2 - r^2)$$

Cone

Figure 1.2.5

See Figure 1.2.5

$$\text{Volume} = \frac{1}{3}\pi r^2 h$$

Pyramid

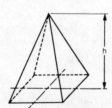

Figure 1.2.6

See Figure 1.2.6.

$$\text{Volume} = \frac{1}{3}\text{ area of base x perpendicular height}$$

Sphere

Figure 1.2.7

See Figure 1.2.7

$$\text{Volume} = \frac{4}{3}\pi r^3$$

$$\text{Surface area} = 4\pi r^2$$

Prism

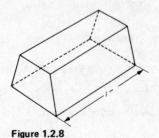

Figure 1.2.8

For any object of constant cross section (Figure 1.2.8)

$$\text{Volume} = \text{area of end x length}$$

Example 1

A clutch disc has an inner and outer radius of 90 mm and 120 mm respectively (Figure 1.2.9). Calculate the area of one face in square metres.

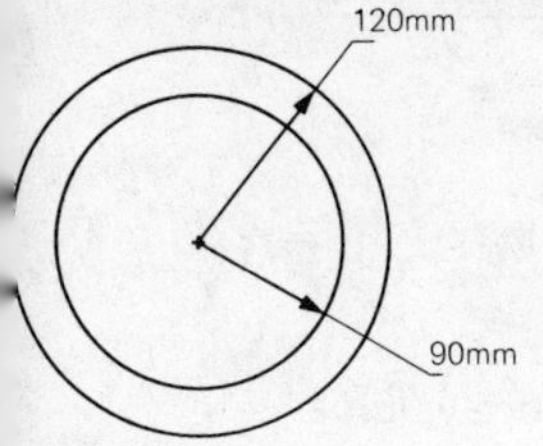

Figure 1.2.9

$$\text{Area} = \text{area of annulus}$$
$$\text{Area} = \pi(R^2 - r^2)$$
$$= \frac{22}{7}(12^2 - 9^2)\ \text{cm}^2$$
$$= \frac{22}{7}(144 - 81)\ \text{cm}^2$$
$$= \frac{22}{\cancel{7}} \times \cancel{63}^{\,9}\ \text{cm}^2$$
$$= 198\ \text{cm}^2$$

There are 100^2 square centimetres in 1 square metre, i.e. $10\,000\ \text{cm}^2 = 1\ \text{m}^2$ or $10^4\ \text{cm}^2 = 1\ \text{m}^2$.

$$\text{Area} = \frac{198}{10^4}$$
$$= 0{\cdot}0198\ \text{m}^2 \quad \text{(decimal point moved four places)}$$

Example 2

A steel ball has a diameter of 16 mm. Calculate its
(a) volume
(b) mass if the density of steel is $0{\cdot}008\ \text{g/mm}^3$.

(a) $\text{Volume of sphere} = \frac{4}{3}\pi r^3$

$$= \frac{4 \times 22 \times 8 \times 8 \times 8}{3 \times 7}$$
$$2145{\cdot}5\ \text{mm}^3$$

(b) $\text{Mass} = \text{volume} \times \text{density}$

$$= 2145{\cdot}5 \times 0{\cdot}008$$
$$= 17{\cdot}164\ \text{g}$$

Example 3

An engine (Figure 1.2.10) has a bore diameter of 80 mm, stroke of 140 mm and a hemispherical combustion chamber. Calculate the
(a) swept volume of the cylinder
(b) clearance volume
(c) compression ratio

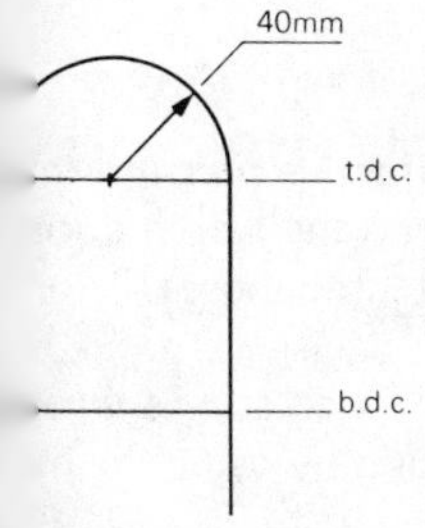

Figure 1.2.10

(a) $\text{Swept volume} = \frac{\pi d^2}{4} \times l$

$$= \frac{22 \times 8 \times \overset{2}{\cancel{8}} \times \overset{2}{\cancel{14}}}{\cancel{7} \times \cancel{4}}\ \text{cm}^3$$
$$= 704\ \text{cm}^3$$

(b)   Volume of hemisphere = Volume of ½ sphere

$$= \frac{\frac{4}{3}\pi r^3}{2} = \frac{4}{6}\pi r^3 = \frac{2}{3}\pi r^3$$

$$= \frac{2 \times 22 \times 4 \times 4 \times 4}{3 \times 7} \text{ cm}^3$$

$$\simeq 134 \text{ cm}^3$$

(c)   Compression ratio $= \dfrac{V_s + V_c}{V_c}$

where V_s = swept volume and V_c = clearance volume

$$= \frac{704 + 134}{134}$$

$$= \frac{838}{134} \simeq 6{\cdot}25$$

Area of an irregular figure

Mid-ordinate rule

The figure is divided into a number of strips of equal width and the average length of each strip is found by measuring the length of the centre line; the mid ordinate (see Figure 1.2.11a).

If h_1 = average height of first strip

then   average height (h) $= \dfrac{h_1 + h_2 + h_3 \dots h_n}{n}$

where n = number of strips.

Area $= h \times b$ (see Figure 1.2.11b)

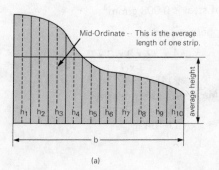

(a)

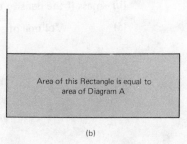

(b)

Figure 1.2.11

This method can be used to find the mean effective pressure (m of an engine (The m.e.p. is the average pressure on the piston durir the power stroke, minus the average pressure which opposes pistor motion during the compression stroke.)

Figure 1.2.12 shows the pressure variation for each stroke. Fror these graphs the average pressures have been found by using the m ordinate rule.

Therefore:

$$\text{m.e.p.} = P_p - P_c$$

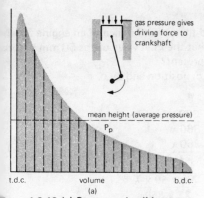

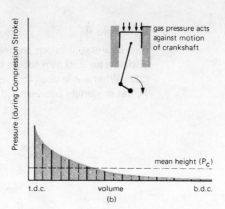

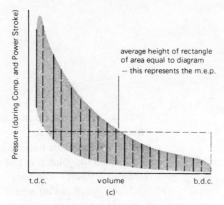

igure 1.2.12 (a) Power stroke, (b) compression stroke, (c) compression and power strokes

where P_p = Average pressure during power stroke

P_c = Average pressure during compression stroke P.

By combining the two graphs used in Figure 1.2.12(a,b), a diagram (Figure 1.2.12c) is obtained which is similar to that produced by an engine indicator. The average length of all the ordinates contained within this diagram is the m.e.p.

ength of an arc

The perimeter or circumference of a circle is πd or $2\pi r$ and this distance represents 360°. Figure 1.2.13 shows that the length of an arc (x) measured around the circumference is given by:

$$\frac{x}{\pi d} = \frac{\theta}{360}$$

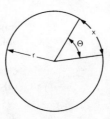

Figure 1.2.13

Example 4

A measurement taken around the circumference of an engine flywh‹ of diameter 350 mm shows that the inlet valve opens 50 mm before‹ t.d.c. What angle does this represent?

Let θ = angle between t.d.c. position and i.v.o.

$$\frac{x}{\pi d} = \frac{\theta}{360}$$

$$\pi d\theta = 360\,x$$

$$\theta = \frac{360\,x}{\pi d}$$

$$\theta = \frac{\overset{180}{\cancel{360}} \times \cancel{50} \times \cancel{x}}{\underset{11}{\cancel{22}} \times \underset{7}{\cancel{350}}}$$

$$\theta \simeq 16°$$

Exercises 1.2

1. A vehicle has an overall length and width of 4415 mm and 1650 mm, respectively. What area does this represent in square metres? (Correct to two decimal places).

2. A rectangular sheet of metal of length 600 mm has an area of 0·264 m². What is the length?

3. A piston has an area of 4400 mm². What is its diameter? Tak‹ π as 22/7.

4. A full drum of oil is tipped into an empty rectangular tank o‹ length 625 mm and width 320 mm. What is the depth of oil in the tank if the cylindrical drum has a diameter of 350 mm and length o‹ 640 mm (take π as 22/7).

5. A gasket with the shape of a parallelogram has two parallel sides of length 45 mm and a perpendicular height of 32 mm. Calcul‹ the area in square centimetres.

6. Two sides of a figure are parallel and are 20 mm apart. If the area is 1200 mm² and the length of one parallel side is 80 mm, find‹ the length of the other parallel side.

7. Calculate the area of a hexagon having a 'distance across flats‹ of 20·8 mm and a length of side of 12 mm.

8. A clutch friction disc has an overall diameter of 220 mm and an inner diameter of 160 mm. Calculate the area of the face in squa‹ centimetres. Take π as 22/7 and give your answer to the nearest whole number.

9. Calculate the total swept volume or cubic capacity of a 6-cylinder engine having a bore of 84 mm and a stroke of 100 mm.

10. A hollow gudgeon pin of length 70 mm has an outer and inner diameter of 22 mm and 16 mm respectively. Calculate the volume of the pin in cubic centimetres.

11. An engine cylinder has a swept volume of 225 cm³ and a clearance volume of 25 cm³.
(a) Calculate the compression ratio.
(b) During a compression test five shots of oil from an oil can decreases the clearance volume by 5 cm³. Calculate the compressio‹ ratio when the oil is in the cylinder.

12. A steel ball has a diameter of 42 mm. Calculate
(a) the volume in cubic millimetres,
(b) the mass if the density if 0·008 g/mm^3.

13. An engine, having a combustion chamber in the form of a true hemisphere, has a stroke of 95 mm. If the distance from b.d.c. to the highest point in the combustion chamber is 130 mm, calculate the diameter of the cylinder bore.

14. How many litres of fuel can be contained in a cylindrical tank of radius 300 mm and length 1260 mm?

15. A tank of constant cross-sectional area of 800 cm^2 has a length of 25 cm. What is the tank capacity in litres?

16. An indicator diagram is divided up into 10 strips of width 12 mm and the mid-ordinates of the strips have lengths in millimetres of 50, 58, 50, 41, 31, 25, 21, 15, 11 and 8. Find the area of the diagram in square millimetres.

17. A measurement taken on the circumference of a flywheel of diameter 420 mm shows that the inlet valve closes 165 mm after b.d.c. What angle does this represent?

18. The static ignition timing is 5° before t.d.c. What distance around the circumference of a crankshaft fan pulley of diameter 140 mm does this angle represent?

19. When checking the automatic advance mechanism with a timing light directed onto the crankshaft fan pulley, it is found that the timing mark moves through a distance of 38 mm measured at a radius of 70 mm on the pulley. What angle does this represent?

20. An engine has a flywheel of diameter 420 mm. During a valve timing check the following distances were measured around the flywheel circumference.

Inlet valve opens	58 mm before t.d.c.
Inlet valve closes	205 mm after b.d.c.
Exhaust valve opens	187 mm before b.d.c.
Exhaust valve closes	77 mm after t.d.c.

(a) Calculate the valve timing to the nearest degree.
(b) Draw a circular type of valve timing diagram for this engine.
(c) Calculate the valve-open-period for each valve.

1.3 Symbols and Formulae

Symbols

Symbols are used to simplify an expression, thereby reducing space and time. In SI, symbols are adopted for unit quantities; the symbols representing the six basic units are m, kg, s, A, cd and K. These forms are universally recognised, but where a symbol is used to represent another quantity, the meaning must be stated, e.g.

$$V_s + V_c = V_t$$

where V_s = cylinder swept volume (cm^3)
V_c = cylinder clearance volume (cm^3)
V_t = cylinder total volume (cm^3)

In this case 'V' is chosen because volume is the quantity. To separate each 'volume' an additional symbol (subscript) is added to the main symbol, e.g. V_s; the 's' enables the term to be identified as the 'swept volume'. Digits are sometimes used to convey a certain meaning, e.g.

$$t = t_1 - t_2$$

where t = change in temperature $^{\circ}$C
t_1 = initial temperature $^{\circ}$C
t_2 = final temperature $^{\circ}$C

The symbol 't_2' should not be confused with t^2. When the number is written as 't^2' it means 't squared' or 't x t'.

Where possible, the symbols used should conform to BS 1991: *Letter Symbols, Signs and Abbreviations.* This recommends the symbol to be used to indicate a given quantity. As there are more quantities that require symbols than letters in the English alphabet, some Greek letters are adopted.

For example, the symbol 't' is recommended for 'time', so if the previous example also involved time, then confusion might occur: to avoid this situation the symbol 'θ' is recommended for 'temperature' so the expression is changed to

$$\theta = \theta_1 - \theta_2$$

When 'time' is introduced into the expression, the meaning of each term should be clear

$$\text{Change of temperature per second} = \frac{\theta_1 - \theta_2}{t}$$

In M.V. work the following Greek letters are used to represent the quantity or unit stated:

Letter		Quantity or unit
α	alpha	coefficient of linear expansion
β	beta	coefficient of cubical expansion; angles
η	eta	efficiency
θ	theta	temperature, angles
μ	mu	coefficient of friction, micro
Ω	omega	ohm (unit of electrical resistance)
π	pi	ratio of circumference of circle to diameter

Formula

A formula is an algebraic statement which shows the relationship between a number of given quantities. An expression written as a formula enables an unknown quantity to be calculated.

Example 1

The compression ratio of an engine is given by

$$R = \frac{V_s + V_c}{V_c}$$

where R = compression ratio
V_s = swept volume of cylinder
V_c = clearance volume of cylinder

If V_s = 300 cm^3 and V_c = 40 cm^3, find the compression ratio.

$$R = \frac{300 + 40}{40}$$

$$R = \frac{340}{40} = 8{\cdot}5$$

Example 2

An engine cylinder has a bore of 70 mm, a swept volume of 300 cm^3 and a clearance volume of 37·5 cm^3.

   (a) What is the compression ratio?

   (b) If the compression height (distance between gudgeon pin centre and crown) is increased by 1 mm, what is the new compression ratio?

    (a)   Original compression ratio $= \dfrac{V_s + V_c}{V_c}$

$$= \frac{300 + 37{\cdot}5}{37{\cdot}5} = 9$$

    (b) Decrease in clearance volume $= \dfrac{\pi d^2}{4} \times l$

$$= \frac{22 \times 7 \times 7 \times 0{\cdot}1}{7 \times 4} \; cm^3$$

$$= 3{\cdot}85 \; cm^3$$

$$\text{New clearance volume} = 37{\cdot}5 - 3{\cdot}85 \; cm^3$$

$$= 33{\cdot}65 \; cm^3$$

$$\text{New compression ratio} = \frac{V_s + V_c}{V_c}$$

$$= \frac{300 + 33{\cdot}65}{33{\cdot}65}$$

$$\simeq 9{\cdot}9$$

    This particular calculation shows that any feature which alters the clearance volume has a considerable effect on the compression ratio: the obvious features are gasket thickness, carbon deposits and sparking plug reach.

    Care must be exercised when using a formula to ensure that the units are suitable. The unit in which the answer is expressed is governed by the units used for each quantity. In the previous example the bore and stroke was given in centimetres, so the unit for the answer can be found by inserting the unit for 'd' and 'l' in the formula, e.g.

$$\frac{\pi d^2}{4} \times l \text{ gives } cm^2 \times cm = cm^3$$

(A very uncommon unit would result if 'd' was expressed in centimetres and 'l' in millimetres.)

The next example shows a similar case

$$V = \frac{1}{3}\pi r^2 h$$

Transposing for h gives

$$h = \frac{3V}{\pi r^2}$$

If V was stated in cubic metres, then 'r' must be expressed in metres to give

$$h = \frac{m^3}{m^2} = \frac{m \times m \times m}{m \times m} = m$$

This shows that units can be cancelled in a similar manner to that used for numerical values.

Example 3

In the formula $\theta = \frac{Q}{mc}$ the quantities are expressed in the following units

Q = joules; m = kilogram; c = joule per kilogram kelvin.

In what unit is θ expressed?
Writing each unit in the formula in symbol form

$$\theta = \frac{J}{kg \times J/kg\ K}$$

J/kg K means $\frac{J}{kg\ K}$

$$\theta = \frac{J}{kg \times \dfrac{J}{kg\ K}}$$

$$\theta = \frac{J \times kg\ K}{kg \times J} = K$$

θ is expressed in 'kelvin' so θ represents temperature

Transposition of formulae

A formula is an equation, so any mathematical operation performed on one side of the equation must also be applied to the other side. Applying this rule to a formula reveals that when a symbol is removed from one side, it appears on the other side with a different sign, e.g.

Transpose $\theta = Q/mc$ to make 'm' the subject.

$$\theta mc = \frac{Qmc}{mc} \quad \text{(multiplying both sides by '}mc\text{')}$$

$$\frac{\theta mc}{\theta c} = \frac{Q}{\theta c} \quad \text{(dividing both sides by '}\theta c\text{')}$$

$$m = \frac{Q}{\theta c}$$

The sign change may be summarised as follows

$$x \rightleftharpoons \div$$

$$\div \rightleftharpoons x$$

$$+ \rightleftharpoons -$$

$$- \rightleftharpoons +$$

square $\rightleftharpoons$ square root

square root $\rightleftharpoons$ square

Example 4

Transpose $V = \pi r^2 h$ to make r the subject.

$$r^2 = \frac{V}{\pi h}$$

$$r = \sqrt{\left(\frac{V}{\pi h}\right)}$$

Example 5

Transpose $V = 4\pi r^3 / 3$ to make r the subject.

$$3V = 4\pi r^3$$

$$r^3 = \frac{3V}{4\pi}$$

$$r = \sqrt[3]{\frac{3V}{4\pi}} \text{ (the cube root of } 3\,V/4\pi)$$

Example 6

Transpose $8 = (V_s + V_c)/V_c$ to make V_c the subject.

$$8\,V_c = V_s + V_c$$

$$8\,V_c - V_c = V_s$$

$$7\,V_c = V_s$$

$$V_c = \frac{V_s}{7}$$

Example 7

The previous example can be applied to the following:
Transpose $R = (V_s + V_c)/V_c$ to make V_c the subject.

$$R V_c = V_s + V_c$$

$$R V_c - V_c = V_s$$

$$V_c (R - 1) = V_s$$

(everything inside the bracket is multiplied by V_c)

$$V_c = \frac{V_s}{R-1}$$

Exercises 1.3

Transpose the following expressions:

1. $A = \dfrac{\pi d^2}{4}$ Make d the subject

2. $A = \dfrac{4}{3} \pi r^3$ Make r the subject

3. $c^2 = a^2 + b^2$ Make a the subject

4. $A = \dfrac{4 + x}{2} \times h$ Make x the subject

5. $A = \pi (R^2 - r^2)$ Make R the subject

6. $R = \dfrac{V_s + V_c}{V_c}$ Make V_s the subject

7. $R = \dfrac{V_s + V_c}{V_c}$ Make V_c the subject

8. $V_c = \dfrac{V_s}{R-1}$ Make R the subject

9. $P = 2 \pi n T$ Make T the subject

10. $T = S p \mu r$ Make p the subject

In questions 11–13, substitute the appropriate units in the formula and establish the unit in which the answer is expressed.

11. $T = S p \mu r$ where S = no units

                            p = newtons

                            μ = no units

                            r = metres

12. $V = \dfrac{\pi d^2}{4} \times l$ d = square metres

                            l = metres

13. $M = \dfrac{4}{3} r^3 p$ r = millimetres

                            p = grammes per cubic millim

14. $S = \dfrac{P}{A}$ P = newtons

                            A = square millimetres

1.4 Graphs

A graph is a diagram which shows how the value of one quantity varies in relation to another. The values are read-off from numbers marked on two lines called *axes*; the horizontal axis is known as the *abscissa* or x-axis, and the vertical axis the *ordinate* or y-axis. Normally the independent quantity is plotted along the abscissa and the dependent quantity (the value obtained by calculation or experiment) is shown as the ordinate.

The scale or spacing of the numbers on the axes should be carefully selected to ensure that all readings can be easily obtained.

The intersection point where the two axes join is called the origin. The origin of each axis, i.e. start of each scale, does not need to be the same number. By selecting a suitable origin for each scale, the space occupied by the graph can be fully utilized.

There are many applications of graphs in M.V. work. The following examples and exercises are intended to give not only practice in using graphs, but also to give an indication of the shape of typical curves associated with our work.

Example 1

Figure 1.4.1 shows the relationship between the linear motion of a piston and the rotary motion of a crankshaft.

(a) Through what angle does the crank rotate from t.d.c. to cause the piston to move a distance equal to half the stroke?

(b) At what point in the stroke is the piston positioned when the crank has moved $90°$ from t.d.c.

(a) Half stroke = 35 mm

   From graph, 35 mm piston movement = crank angle of $84°$.
(b) From graph:

   Crank angle of $90°$ = piston movement of 38 mm.

This example shows that the piston moves more than half the stroke when the crank rotates through $90°$ from t.d.c.

The line diagram of the piston and crank which gives the piston movement for a given crank movement is called a piston displacement diagram.

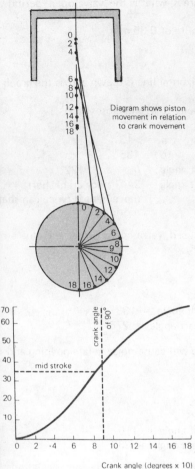

Diagram shows piston movement in relation to crank movement

Crank angle (degrees x 10)

Figure 1.4.1

Example 2

Figure 1.4.2 shows a graph of cam displacement (lift of a valve which has no clearance) against camshaft rotation for a modern engine.

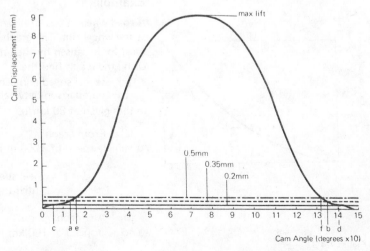

Figure 1.4.2

Expressed in crankshaft degrees, what is the 'valve open' period when the valve clearance is set to
  (a) the recommended figure of 0·35 mm
  (b) 0·2 mm
  (c) 0·5 mm

To find the period, a horizontal line is drawn across the graph at the appropriate clearance.
  (a) From graph, valve opens at '*a*' and closes at '*b*'.

$$a = 13°$$
$$b = 135°$$

∴ camshaft angle $= b - a = 122°$
and crankshaft angle $= 244°$ since crankshaft movement is twice camshaft motion.

(b) When clearance is reduced, valve opens earlier and closes later.
From graph

$$c = 5°$$
$$d = 143°$$
$$\text{camshaft angle} = 138°$$
$$\text{crankshaft angle} = 276°$$

(c) An excessive clearance will cause noise, a late opening and an early closing of the valve.
From graph

$$e = 16°$$
$$f = 132°$$
$$\text{camshaft angle} = 116°$$
$$\text{crankshaft angle} = 232°$$

This example shows that with a modern engine, any deviation from the recommended valve clearance can cause an alteration in engine performance.

Example 3

A road wheel is out-of-balance to the extent of 100 grams, measured at the wheel rim. The graph shown in Figure 1.4.3 shows the centrifugal force caused by the unbalanced mass when the wheel is rotated. Using the graph find the
  (a) wheel speed and centrifugal force at a vehicle speed of 100
  (b) comparison between the force at a vehicle speed of 160 km
to that given at 80 km/h.

(a) From graph:
At vehicle speed of 100 km/h.

$$\text{wheel speed} = 12·5 \text{ rev/s}$$
$$\text{centrifugal force} = 218 \text{ N}$$

(b) Also from graph:
At vehicle speed of 160 km/h

$$\text{force} = 560 \text{ N}$$

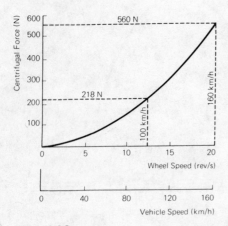

Figure 1.4.3

At vehicle speed of 80 km/h

$$\text{force} = 140 \text{ N}$$

So the force at 160 km/h is four times that at 80 km/h.

This example shows that centrifugal force increases with the square of the speed. The high rate of increase of the force in respect to speed shows that, although the effect of the out-of-balance of a rotating member may be small at low speeds, the effect at high speeds is considerable.

Some graphs show a number of curves which can be 'read-off' from one or more scales. This type of graph is shown in the next example.

Example 4

Figure 1.4.4 shows a graph of brake performance. Using this graph determine the

(a) stopping distance, from a speed of 50 km/h if the brake efficiency is 60 per cent, and

(b) brake efficiency if the vehicle is brought to rest in 40 m from a speed of 72 km/h.

(a) The ordinate drawn at 50 km/h cuts the curve representing 60 per cent at a point shown on the scale as 16 m.

(b) The intersection of the ordinate and abscissa occurs at an efficiency of 50 per cent.

This graph shows that for a given brake efficiency, the stopping distance increases with the square of the road speed: a vehicle travelling at a speed of 60 km/h requires a stopping distance four times as great as that needed for 30 km/h.

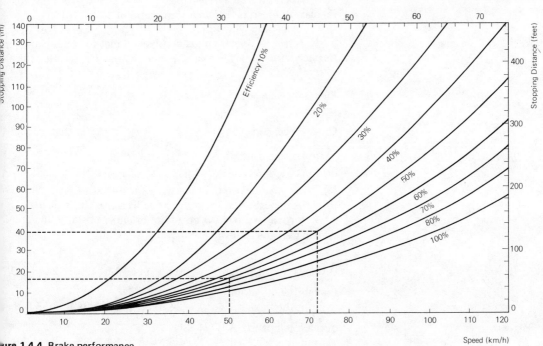

Figure 1.4.4 Brake performance

Exercises 1.4

1. The torque output in relation to speed of a 4-cylinder engine

Speed (rev/min)	1000	1500	2000	2500	3000	3500	4000	4500	
Torque (Nm)	50	56	60	62	61	58	53	48	

Draw a smooth curve through these points and determine
(a) torque at a speed of 3750 rev/min
(b) speed at which maximum torque is developed

2. The specific fuel consumption of a 12-cylinder high performance engine is given by the table:

Speed (rev/min)	1000	1500	2000	2500	3000	3500	4000	4500	5000	5500	6000	65
Specific fuel consumption (litre/kW h)	0·40	0·39	0·38	0·40	0·39	0·38	0·37	0·37	0·38	0·40	0·42	0·

Plot the graph and determine the range of speed where the engine most economical, i.e. the lowest part of the curve.

3. Analysis of the exhaust gas discharged from a petrol engine shows that the gas composition varies with the air/fuel ratio. The following table shows the percentage of each gas.

Air/fuel ratio	11	12	13	14	15	16	17	18	19
CO_2 (%)	4	7	9	12	15	14	13	12	11
CO (%)	16	11	7	4	0·5	0	0	0	0
O_2 (%)	0	0	0	0	0	1	3	4	5

Plot these results on a base of 'air/fuel ratio' and determine the
(a) gas composition for an air/fuel ratio of 12·5 : 1
(b) air/fuel ratio which produces the largest proportion of CO_2

4. A test on a steering system showed that the outer wheel moved through a smaller angle than the inner wheel. The results of the test were:

Inner wheel angle (degrees)	5	10	15	20	25	30
Outer wheel angle (degrees)	4	8·5	13	17	21	24

Plot these results and determine the angle of the
(a) outer wheel when the inner wheel is turned through 26°
(b) inner wheel when the outer wheel is turned through 10°

5. A road test to show the effect of a thermostat in the cooling system gave the following results:

Distance travelled (km)	0·5	1	1·5	2·0	2·5	3·0
Coolant temperature without thermostat (°C)	36	38·5	39	39·5	40·5	44
Coolant temperature with thermostat (°C)	55	76	84	85	83	85

Plot the two curves and determine the temperature:
(a) difference (at a distance of 1·75 km) between an engine with a thermostat and one without a thermostat
(b) of an engine with a thermostat after travelling a distance of 0·75 km
   6. The mechanical advance given in distributor degrees for an engine is shown in the following table.

Distributor speed (rev/min)	500	750	1000	1250	1500	1750	2000	2250	2500	2750	3000
Maximum advance (degrees)	0	3·5		7	8	8·5		9	9·5	9·75	10	10	10
Minimum advance (degrees)	0	1·5		5	6	6·5		6·75	7·25	7·5	8	8	8

Plot the two curves.
Bearing in mind that the crankshaft rotates at twice the speed of the distributor, determine the:
(a) maximum advance in distributor degrees for a crankshaft speed of 2600 rev/min
(b) minimum advance in crankshaft degrees for a crankshaft speed of 5600 rev/min
   7. Draw a straight line graph of 'distance in kilometres' against 'distance in miles'. The scale should range from 0 to 100 miles.
Using the graph convert
(a) 50 miles to kilometres
(b) 87 miles to kilometres
(c) 30 mile/h to km/h
(d) 70 mile/h to km/h
   8. The vacuum advance expressed in distributor degrees for an engine is shown in the following table.

Vacuum (mm Hg)	100	125	150	175	200	225	250	275
Maximum advance (degrees)	0·75	1·75	3·75	5·75	7·5	8·5	8·5	8·5
Minimum advance (degrees)	0	0	1	2·75	4·5	5·5	5·5	5·5

Plot these curves with the 'vacuum' scale on the x-axis.
(a) What is the minimum distributor advance which should be given by a 'vacuum' of 180 mm of mercury?
(b) What 'vacuum' is required to produce a maximum advance of 7 distributor degrees?

5 Measurement

e vernier caliper

This instrument enables a wide range of linear measurements to be determined. It gets its name from the type of scale used. Figure 1.5.1 shows that the instrument consists of two main parts
   1. the fixed jaw or body on to which is engraved the main scale
   2. the sliding jaw which contains the vernier scale

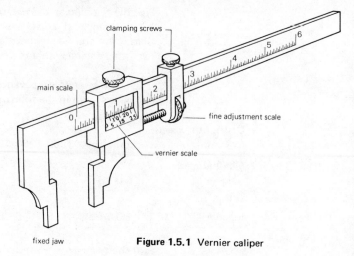

Figure 1.5.1 Vernier caliper

Figure 1.5.2

Positioned adjacent to the main scale, the vernier scale divisions are slightly smaller than those used for the main scale (Figure 1.5. shows an enlargement of the scale).

This arrangement uses a vernier scale having 25 divisions and these are spaced over a main scale length of 12 mm. Each division on the vernier scale equals

$$\frac{1}{25} \text{ of } 12 \text{ mm} = 0.48 \text{ mm}$$

When the jaws of the instrument are opened and the first division on the vernier scale lines up with the first division on the main sca the distance moved by the jaw is equal to the differences in size between the division on the main scale and the division on the ver scale. The small divisions on the main scale represent 0.5 mm, so

$$\text{distance moved} = 0.50 - 0.48 \text{ mm} = 0.02 \text{ mm}$$

So each vernier division represents 0.02 mm
The measurement is obtained by adding to the main scale reading the value indicated by the line on the sliding scale which is level w a division mark on the main scale.

Example 1

Reading (see Figure 1.5.3)

top scale	= 3.5
bottom scale (15 x 0.02)	= 0.30
total	= 3.8 mm

Figure 1.5.3

Example 2

Reading (see Figure 1.5.4)

top scale	= 27.5
bottom scale (7 x 0.02)	= 0.14
total	27.64 mm

Figure 1.5.4

ther vernier scales

Not all verniers use 25 divisions on the scale; some use 50 and others 10, but the principle of each type is the same.

Scale with 50 divisions: The main scale has divisions of 1 mm, and the vernier scale spreads over 49 mm.

$$1 \text{ vernier scale division} = \frac{1}{50} \text{ of 49 mm}$$

$$= \frac{49}{50} = 0.98 \text{ mm}$$

$$\binom{\text{Distance moved by jaw from zero to}}{\text{line-up first vernier scale division}} = \binom{\text{difference between top}}{\text{and bottom divisions}}$$

$$= 1.00 - 0.98$$
$$= 0.02 \text{ mm}$$

One division on vernier scale represents 0.02 mm.

Scale with 10 divisions: The main scale has divisions of 1 mm and the vernier scale spreads over 9 mm.

$$1 \text{ vernier scale division} = \frac{1}{10} \text{ of 9 mm}$$

$$= \frac{9}{10} = 0.9 \text{ mm}$$

$$\binom{\text{Distance moved by jaw from zero to}}{\text{line-up first vernier scale division}} = 1.0 - 0.9$$

$$= 0.1 \text{ mm}$$

One division on vernier scale represents 0.1 mm.

ternal measurements

By reducing the ends of the jaws to a given width, internal measurements can be taken. This width is normally stated on the jaw and must be added to the scale readings.

ptical measuring equipment

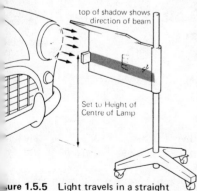

Optical methods are sometimes used for checking steering geometry and headlamp alignment. Although some equipment is quite elaborate, the operating principles are based on one or two simple laws of light. *Light travels in a straight line.* Examination of a shadow shows that a ray of light travels in a straight line.

This feature can be used to check headlamp alignment (Figure 1.5.5). After setting the equipment to a height equal to the distance between the ground and the headlamp centre, the equipment is moved to a position about 0.5 metre in front of the lamp. By observing the position of the edge of the shadow the vertical setting of the lamp can be determined.

The angle of reflection from a polished surface is equal to the angle at which the light is applied, i.e. the angle of reflection is equal to the angle of incidence.

ure 1.5.5 Light travels in a straight line—this feature is used to check headlamp position

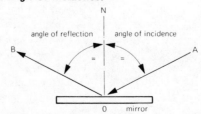

Figure 1.5.6

A plane (flat) mirror can be used to show this law (Figure 1.5.6)
The angle AON is always equal to angle BON.

One type of wheel alignment gauge uses this principle (Figure
1.5.7). The mechanic looks through the eye piece and rotates the
screen slightly until a line is centred in the reflected image of the
screen pattern. Toe-in or toe-out is read-off from a scale attached t
the screen.

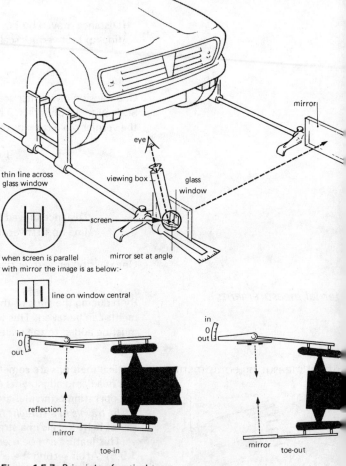

Figure 1.5.7 Principle of optical type
alignment gauge

Exercises 1.5

1. A vernier scale has 10 divisions spaced over a main scale len
of 9 mm. What distance is represented by each division on the ver
scale?

2. A vernier scale has 25 divisions spaced over a main scale len
of 12 mm. What distance is represented by each division on the ve
scale?

3. A vernier scale had 50 divisions spaced over a main scale len
of 49 mm. What distance is represented by each division on the ve
scale?

Questions 4–6 refer to a vernier scale having 10 divisions. State t
reading in each case.

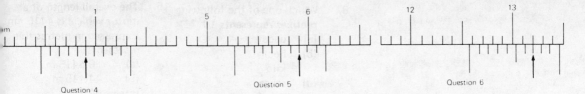

Question 4   Question 5   Question 6

Questions 7—19 refer to a vernier scale having 25 divisions. State the reading in each case.

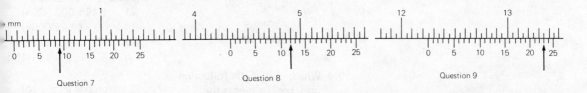

Question 7   Question 8   Question 9

Questions 10—12 refer to a vernier scale having 50 divisions. State the reading in each case.

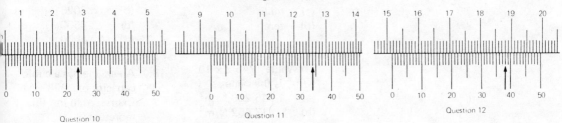

Question 10   Question 11   Question 12

6 Multiple choice questions

1. The metre is the unit of
   - (a) length
   - (b) mass
   - (c) capacity
   - (d) area

2. The kilogram is the unit of
   - (a) force
   - (b) capacity
   - (c) energy
   - (d) mass

3. The candela is the unit of
   - (a) light intensity
   - (b) electrical current
   - (c) energy
   - (d) power

4. The kelvin is the unit of
   - (a) power
   - (b) light
   - (c) heat
   - (d) temperature

5. The symbol Mm represents
   - (a) mega-metres
   - (b) milli-metres
   - (c) micro-metres
   - (d) multi-metres

6. The prefix 'kilo' means
   - (a) 1 000 000
   - (b) 1000
   - (c) 100
   - (d) 10

7. A micro-ampere is
   - (a) one million amperes
   - (b) one thousand amperes
   - (c) one thousandth part of an ampere
   - (d) one millionth part of an ampere

8. Which one of the following prefixes represents 10^3 ?
   (a) deca
   (b) hecto
   (c) kilo
   (d) mega

9. Which one of the following prefixes represents the smallest value?
   (a) micro
   (b) mega
   (c) deca
   (d) giga

10. Which one of the following prefixes represents the largest value?
    (a) nano
    (b) kilo
    (c) milli
    (d) tera

11. The value $1745 \cdot 2 \times 10^3$ N/m^2 is the same as
    (a) $1\,745\,200\ N/m^2$
    (b) $17\,452\ N/m^2$
    (c) $174 \cdot 52\ N/m^2$
    (d) $1 \cdot 7452\ N/m^2$

12. The values $655 \times 10^6\ N/m^2$ is the same as
    (a) $655\,000\,000\ N/m^2$
    (b) $655\,000\ N/m^2$
    (c) $0 \cdot 655\ N/m^2$
    (d) $0 \cdot 000\,655\ N/m^2$

13. How many millimetres equal 1 kilometre?
    (a) 100 or 10^2
    (b) 1000 or 10^3
    (c) 10 000 or 10^4
    (d) 1 000 000 or 10^6

14. What fraction of a kilogram is a milligram?
    (a) $0 \cdot 100$ or 10^{-1}
    (b) $0 \cdot 001$ or 10^{-3}
    (c) $0 \cdot 000\,001$ or 10^{-6}
    (d) $0 \cdot 000\,000\,1$ or 10^{-7}

15. The overall length of a motor vehicle is 4415 mm Expressed in metres this represents
    (a) $0 \cdot 4415$ m
    (b) $4 \cdot 415$ m
    (c) $44 \cdot 15$ m
    (d) $441 \cdot 5$ m

16. The mass of a motor vehic is 1195 kg. Expressed in grams this represents
    (a) 1 195 000 000 g
    (b) 1 195 000 g
    (c) $119 \cdot 5$ g
    (d) $1 \cdot 195$

17. A piston has an area of $45\ cm^2$. Expressed in mm this represents
    (a) $0 \cdot 45$
    (b) 2025
    (c) 4500
    (d) 202 500

18. An engine has a cubic capacity of $1998\ cm^3$ Expressed in m^3 this represents
    (a) $0 \cdot 001\,998$ or $1 \cdot 998 \times 10^{-3}$
    (b) 666
    (c) 5994
    (d) 1 998 000 or $1 \cdot 998 \times 10^6$

19. Given that $1\ dm^3 = 1$ litre a petrol tank of volume $63\,600\ cm^3$ has a capacit of
    (a) $21 \cdot 2$ litres
    (b) $63 \cdot 6$ litres
    (c) 636 litres
    (d) 2120 litres

20. An engine has a cubic capacity of $0 \cdot 002\,496\ m^3$ Expressed in litres this represents
    (a) $0 \cdot 2496$
    (b) $0 \cdot 632$
    (c) $2 \cdot 496$
    (d) $6 \cdot 32$

21. A vehicle has an overall length and width of 4778 mm and 1695 mm respectively. Expressed in square metres and corrected to two decimal places, the vehicle occupies a garage floor area of
    (a) 8·09
    (b) 8·1
    (c) 8·11
    (d) 8·16

22. A bolt has a cross-sectional area of 38·5 mm^2. If π is taken as 22/7 the bolt diameter is
    (a) 0·14 mm
    (b) 3·5 mm
    (c) 7 mm
    (d) 49 mm

23. A cube has a side length of 2 cm. The distance from the lower corner to the upper opposite corner is
    (a) 2·828 cm
    (b) 3·464 cm
    (c) 8 cm
    (d) 12 cm

24. An engine has a stroke of 8 cm and a connecting-rod length of 16 cm. When the connecting-rod forms a right-angle with the crank throw, the distance between the centre of the main bearing and the centre of the gudgeon-pin is
    (a) 15·49 cm
    (b) 16·49 cm
    (c) 17·89 cm
    (d) 19·29 cm

25. A triangle has an area of 0·0494 m^2. If the length of the base is 380 mm, the perpendicular height in millimetres is
    (a) 1·3
    (b) 26
    (c) 130
    (d) 260

26. A four-sided figure which has only two sides parallel is called a
    (a) square
    (b) cube
    (c) parallelogram
    (d) trapezium

27. The area of one face of a cube of volume 64 mm^3 is
    (a) 4 mm^2
    (b) 8 mm^2
    (c) 14 mm^2
    (d) 16 mm^2

28. Which one of the following is a unit of volume?
    (a) mm
    (b) cm^2
    (c) m^3
    (d) litre

29. The area of an annulus is given by
    (a) $\frac{4}{3}\pi r^3$
    (b) $\frac{1}{3}\pi r^2 h$
    (c) $\frac{2}{3}\pi (R - r)$
    (d) $\pi (R^2 - r^2)$

30. The volume of a sphere is given by
    (a) πr^2
    (b) $\frac{1}{3}\pi r^2$
    (c) $\frac{2}{3}\pi r^3$
    (d) $\frac{4}{3}\pi r^3$

31. The volume of a hemisphere is given by
    (a) $\frac{2}{3}\pi r^3$
    (b) $\frac{4}{3}\pi r^3$
    (c) $\frac{1}{3}\pi r^3$
    (d) $\frac{1}{3}\pi r^2$

32. The volume of a cone is given by
    (a) $\pi r^2 h$
    (b) $\frac{1}{3}\pi r^2 h$
    (c) $\frac{2}{3}\pi r h$
    (d) $\frac{2}{3}\pi r^3 h$

33. The surface area of a sphere is given by
    (a) $\frac{1}{3}\pi r$
    (b) $\frac{1}{3}\pi r^2$
    (c) $4\pi r^2$
    (d) $\frac{2}{3}\pi r^3$

34. What volume is occupied by 5 ml of oil?
    (a) $0.05\ cm^3$
    (b) $0.5\ cm^3$
    (c) $5\ cm^3$
    (d) $50\ cm^3$

35. The mid-ordinate rule is a method for finding the
    (a) area of an irregular figure
    (b) length of sides in a triangle
    (c) mass when the volume is known
    (d) length of an arc

36. A circle of diameter 100 mm has an area of 7854 mm^2, whereas when the diameter is 80 mm the area is 5027 mm^2. Using these values, the area of an annulus of outer diameter 100 mm and inner diameter 80 mm is
    (a) $2827\ mm^2$
    (b) $5027\ mm^2$
    (c) $7854\ mm^2$
    (d) $12\ 881\ mm^2$

37. A 4-cylinder engine has a bore area of 3600 mm^2 and a stroke of 70 mm. The total swept volume (cubic capacity) is
    (a) $252\ cm^3$
    (b) $1008\ cm^3$
    (c) $2520\ cm^3$
    (d) $1\ 008\ 000\ cm^3$

38. A 4-cylinder engine of capacity 2 litres has a clearance volume of 62·5 cm^3 per cylinder. The compression ratio is
    (a) 5 : 1
    (b) 7 : 1
    (c) 8 : 1
    (d) 9 : 1

39. The swept volume of a cylinder of bore 80 mm and stroke 70 mm is
    (a) $50\ cm^3$
    (b) $56\ cm^3$
    (c) $352\ cm^3$
    (d) $5600\ cm^3$

40. A tank of constant trian cross-section has a lengt 300 mm. If the cross-sec has a base of 200 mm ar vertical height of 400 m the volume of the tank i
    (a) $400\ cm^3$
    (b) $6000\ cm^3$
    (c) $12\ 000\ cm^3$
    (d) $24\ 000\ cm^3$

41. The capacity in litres of rectangular tank of leng 80 cm, breadth 40 cm a height 60 cm is
    (a) 0·192
    (b) 19·2
    (c) 192
    (d) 192 000

42. A cylindrical hose has internal and external diameters of 50 mm and 63 mm respectively. If the internal circumference is 157 mm, the external circumference is
    (a)   13 mm
    (b)   44 mm
    (c)   170 mm
    (d)   198 mm

43. A hemisphere of diameter 42 mm has a volume of
    (a)   9·24 cm^3
    (b)   19·404 cm^3
    (c)   38·808 cm^3
    (d)   155·232 cm^3

44. Which one of the following volumes represents a capacity of 1 litre?
    (a)   1 mm^3
    (b)   1 cm^3
    (c)   1 dm^3
    (d)   1 m^3

45. The volume of a tank is 0·5 m^3; this represents a capacity in litres of
    (a)   5
    (b)   50
    (c)   500
    (d)   5000

46. The area of a figure is found by multiplying the average length of mid-ordinate by the
    (a)   distance between each ordinate
    (b)   number of ordinates
    (c)   base of the figure
    (d)   height of the figure

47. An irregular figure is divided up into 12 equal strips of width 18 mm. If the average length of each strip is 60 mm, the total area of the figure is
    (a)   216 mm^2
    (b)   720 mm^2
    (c)   1080 mm^2
    (d)   12 960 mm^2

48. A crankshaft fan pulley of diameter 140 mm has a timing mark 11 mm from t.d.c. measured around the circumference. This distance represents an angle of
    (a)   4°
    (b)   7°
    (c)   9°
    (d)   11°

49. A flywheel ring gear has 144 teeth. Moving the crankshaft through a distance of 12 teeth represents an angle of
    (a)   8°
    (b)   12°
    (c)   15°
    (d)   30°

50. A measurement taken around the circumference of an engine flywheel of diameter 350 mm shows that the inlet valve opens 55 mm before t.d.c. This distance represents an angle of
    (a)   5·5°
    (b)   17°
    (c)   18°
    (d)   56°

51. Transpose $c^2 = a^2 + b^2$ to make 'b' the subject.
    (a) $b = c - a$
    (b) $b = \sqrt{(c^2 - a^2)}$
    (c) $b = \sqrt{(c^2 + a^2)}$
    (d) $b = c/a$

52. Transpose $a = \dfrac{bc}{d} + e$ to make '*b*' the subject.

(a) $b = \dfrac{ad}{c} - e$

(b) $b = \dfrac{c}{ad - e}$

(c) $b = \dfrac{ad - e}{c}$

(d) $b = \dfrac{d(a - e)}{c}$

53. Transpose $V = \dfrac{4}{3}\pi r^3$ to make '*r*' the subject.

(a) $r = \sqrt[3]{\left(\dfrac{3V}{4\pi}\right)}$

(b) $r = \sqrt[3]{\left(\dfrac{4\pi}{3V}\right)}$

(c) $r = \sqrt[3]{\left(\dfrac{V}{\frac{3}{4}\pi}\right)}$

(d) $r = \sqrt[3]{\left(V - \dfrac{4}{3}\pi\right)}$

54. Transpose $T = S p \mu r$ to make '*r*' the subject.

(a) $r = \dfrac{Sp\mu}{T}$

(b) $r = \dfrac{T}{Sp\mu}$

(c) $r = \dfrac{ST}{p\mu}$

(d) $r = T - Sp\mu$

55. Transpose $V = \dfrac{\pi}{4}(D^2 - d^2)$ to make '*D*' the subject.

(a) $D = \sqrt{\left(\dfrac{4V}{\pi} + d^2\right)}$

(b) $D = \sqrt{\left(\dfrac{\pi V}{4} + d^2\right)}$

(c) $D = \sqrt{\left(V - \dfrac{\pi}{4} + d^2\right)}$

(d) $D = \dfrac{V}{2} - \dfrac{\pi}{2} + d$

56. The '*y*' axis of a graph is called the
(a) origin
(b) abscissa
(c) ordinate
(d) horizontal axis

57. The '*x*' axis of a graph is called the
(a) origin
(b) abscissa
(c) ordinate
(d) vertical axis

58. The intersection point where the two axes join is called the
(a) origin
(b) abscissa
(c) ordinate
(d) scale

59. A vernier scale has 25 divisions spaced over a main scale length of 12 mm. When the jaws are opened to cause the first division line on the vernier scale to line up with the second line on the main scale, the reading is
(a) 0·1 mm
(b) 0·2 mm
(c) 0·01 mm
(d) 0·02 mm

60. A vernier caliper, having a width of jaws of 10 mm is used to check an internal measurement. If the bore being measured is 63·25 mm the reading on the scale is
(a) 0·25 mm
(b) 53·25 mm
(c) 63·25 mm
(d) 73·25 mm

61. A vernier scale has 50 divisions spaced over a main scale length of 49 mm. When the jaws are opened to line up the first division line on the vernier scale with the second line on the main scale, the reading is
(a) 0·1 mm
(b) 0·2 mm
(c) 0·01 mm
(d) 0·02 mm

.1 Conversion of eat

uantity of heat

eat capacity

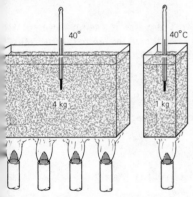

40°   40°C

4 kg   1 kg

ure 2.1.1 To achieve the same final temperature a mass of 4 kg requires four times the heat required by a mass of 1 kg

ecific heat capacity

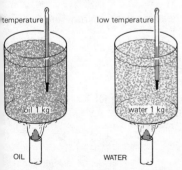

temperature   low temperature

oil 1 kg   water 1 kg

OIL   WATER

re 2.1.2 When equal quantities of oil and water are heated, the oil temperature will rise at a faster rate — the specific heat capacity of water is greater

Heat is a form of energy. To raise the temperature of a substance, energy must be given to it. When we speak about heat being supplied to an object, we mean the amount of energy being delivered. Since the unit of energy is the joule, then

Quantity of heat is the energy in joules

Temperature and heat are two independent terms which should not be confused. It will be remembered that temperature is the 'degree of hotness' of a body. Figure 2.1.1 shows an example in which two materials have a similar 'hotness', i.e. temperature, yet the heat contained in each one is very different. The larger mass takes four times as much heat to raise its temperature a given amount, but after heating contains four times the energy of the smaller mass. In this example, the mass of 4 kg is said to have a larger heat capacity than the 1 kg mass.

Heat capacity of a body is the heat required to raise its temperature by 1°C

The SI unit of heat capacity is the joule per degree C (J/°C or J/K).

The heat capacity varies with substances. If equal masses of oil and water are heated as shown in Figure 2.1.2, the oil temperature will rise at a faster rate. This indicates that the water requires more heat to raise its temperature a given amount, i.e. the water has a higher heat capacity.

The heat capacity of a substance can be found by consulting a table. When a mass of 1 kg is used as a standard the heating value is called the *specific heat capacity*.

Specific heat capacity is the heat required to raise one kilogram of a substance through 1°C (J/kg°C or J/kg K)

Substance	Specific heat capacity (J/kg K)
Water	4200
Ice	2100
Oil	1700
Aluminium	900
Steel	480
Iron	460
Copper	400
Brass	380
Zinc	380
Lead	130

The table of specific heat capacities shows that water has a very high value and this property makes water very attractive to use as cooling or quenching media.

Example 1

A cooling system contains 5 litres of water at a temperature of 16 How much heat is required to raise the water temperature to 96 $^\circ$ (Specific heat capacity of water = 4200 J/kg k)

Mass of 1 litre of water = 1 kg
Heat energy = Mass x specific heat capacity x temperature change
$$= 5 \times 4200 \times (96 - 16)$$
$$= 5 \times 4200 \times 80$$
$$= 1\ 680\ 000\ J$$
$$= 1680\ kJ\ (OR\ 1.68\ MJ)$$

Example 2

A block of copper of mass 2 kg has a temperature of 20 $^\circ$C. What be the temperature of the block after it has received heat energy o 240 kJ. (Specific heat capacity of copper = 400 J/kg K).

Heat received = Mass x specific heat capacity x temperature change
$$Q = m \times c \times (t_2 - t_1)$$

OR

where
Q = quantity of heat in J
m = mass in kg
c = specific heat capacity in J/kg K
t_1 = initial temperature in $^\circ$C
t_2 = final temperature in $^\circ$C

Transposing for t_2

$$t_2 - t_1 = \frac{Q}{m \times c}$$

$$t_2 = \frac{Q}{m \times c} + t_1$$

Substituting values

$$t_2 = \frac{240\,000}{2 \times 400} + 20$$

$$t_2 = \frac{240\,000}{800} + 20$$

$$t_2 = 300 + 20 = 320\,^\circ C$$

Heat exchange

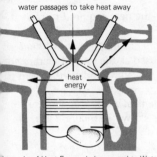

water passages to take heat away

heat energy

Example of Heat Energy being passed to Water

hot

oil or water

temperature of liquid increases

heat energy given up to liquid

Figure 2.1.3 Heat exchange

On motor vehicles there are many cases of heat being exchanged from one substance to another. In all cases the heat transfer takes place from a hot to a cold substance; the transfer follows the law that

heat lost = heat gained

Figure 2.1.3 shows a simple example where a hot object is quenched in a cold liquid. This method could be used to find the specific heat capacity of a substance. The liquid container, or calorimeter as it is often called in experimental work, should be insulated against heat loss to the atmosphere if accurate results are required.

Example 3

A block of steel of mass 2 kg at 800 $^\circ$C is plunged into a mass of 4 kg of water at 15 $^\circ$C. If heat losses are neglected, what is the final temperature of the steel and water? The specific heat capacities of water and steel can be taken as 4·2 kJ/kg K and 0·48 kJ/kg K, respectively.

Let the final temperature of the steel and water = t
Heat lost by the steel = mass x specific heat capacity
x temperature change
= 2 x 480 x (800 − t)
Heat gained by the water = mass x specific heat capacity
x temperature change
= 4 x 4200 x (t − 15)
heat lost = heat gained
2 x 480 x (800 − t) = 4 x 4200 x (t − 15)
960 (800 − t) = 16 800 (t − 15)

Multiplying everything in the bracket by the number outside

$$768\,000 - 960t = 16\,800t - 252\,000$$
$$768\,000 + 252\,000 = 16\,800t + 960t$$
$$1\,020\,000 = 17\,760t$$

$$\therefore \quad t = \frac{1\,020\,000}{17\,760} \simeq 57^\circ C$$

Energy conversion

You are aware that if you rub your hands together, heat is generated. The work done overcoming friction is converted into heat and this warms your hands.

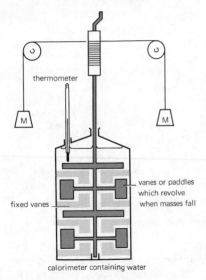

thermometer

vanes or paddles
which revolve
when masses fall

fixed vanes

calorimeter containing water

Figure 2.1.4 Joule's apparatus

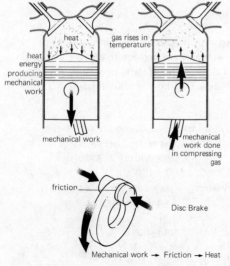

heat

gas rises in
temperature

heat
energy
producing
mechanical
work

mechanical work

mechanical
work done
in compressing
gas

friction

Disc Brake

Mechanical work → Friction → Heat

Figure 2.1.5 Conversion of energy

The relationship between mechanical units of work and heat energy was first proved by James Joule with an apparatus similar to Figure 2.1.4. His experiments showed that energy given to water by falling masses caused the water temperature to increase. Today we take this energy conversion for granted, but in order to honour James Joule for his achievements in showing that mechanical, electrical and chemical energy could all be converted into heat, his name is used the unit for energy.

With the Imperial system a conversion factor is necessary in order to change mechanical units of work to heat energy and vice versa. This factor is given as

$$778 \text{ ft lbf of mechanical work} = 1 \text{ Btu}$$

(Btu stands for British thermal unit—the heat required to raise the temperature of 1 lb of water by 1 °F)

This relationship is called the mechanical equivalent of heat, or Joule's equivalent. In SI the joule is used as the unit of heat energy as well as the unit of mechanical energy, so nowadays the term 'mechanical equivalent of heat' has no use.

There are many cases where mechanical work is converted to heat and vice versa; some of these are shown in Figure 2.1.5.

Example 4

To stop a vehicle, the energy of motion has to be converted into heat by the brakes. What temperature will be reached by a brake disc is an energy of 69 kJ is converted by each brake? (Assume all of the heat is transferred to the disc and no heat is lost to the air.)

Mass of disc = 3 kg
Specific heat capacity of disc = 460 J/kg K
Initial temperature   = 20 °C
Mechanical energy = heat energy = 69 000 J
Heat gained by disc = mass x specific heat capacity
                       x temperature change
$$69\,000 = 3 \times 460 \times (t - 20)$$

where t = final temperature of disc in °C

$$t - 20 = \frac{69\,000}{3 \times 460}$$

$$t = \frac{69\,000}{3 \times 460} + 20$$

Final temperature of disc = 50 + 20 = 70 °C

In practice the temperature would be less than 70 °C because heat would also be transferred to
1. the air by radiation,
2. the caliper assembly which in turn will heat the fluid, and
3. the hub by conduction. It is for this reason that special grease having a high melting point is used in the hubs of vehicles having brakes.

alorific value of a fuel

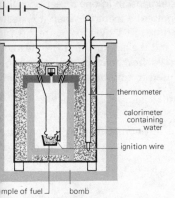

thermometer

calorimeter
containing
water

ignition wire

mple of fuel   bomb

igure 2.1.6 Bomb calorimeter — amount
that the water is heated
depends on the heat given out
by the fuel

When a fuel is burnt, heat is generated. The quantity of heat liberated
is determined by burning a given mass of fuel in a special calorimeter
(Figure 2.1.6) and measuring the heat given out. The value obtained
is called the calorific value.

**Calorific value of a fuel is a measure of the heat units
contained in a given mass of fuel**

Typical values are

Petrol   44 MJ/kg (44 megajoule per kilogram)
Diesel fuel 42 MJ/kg

The variation in calorific values between different petrols is very
small, so for practical purposes the very small difference between the
various grades does not justify further consideration. Calorific value
of a fuel is used when the thermal efficiency of an engine is to be
calculated.

xercises 2.1

1. Find the heat capacity of a mass of 5 kg of copper. (Specific
heat capacity of copper = 400 J/kg K).

2. The heat capacity of a mass of aluminium is 7·2 kJ/K. What
is the mass? (Specific heat capacity of aluminium = 900 J/kg K).

3. How much heat energy is required to raise a mass of 20 kg
from 16 °C to 56 °C. (Specific heat capacity = 480 J/kg K).

4. Water enters the engine at 65 °C and leaves at 85 °C. How
much heat energy is lost to the cooling water per minute if the flow
rate is 10 kg/min.

5. A block of steel of mass 2 kg has a temperature of 10 °C. What
will be the temperature of the block after it has received heat energy
of 240 kJ? (Specific heat capacity of steel = 480 J/kg K).

6. A mass of 5 kg of water has a temperature of 10 °C. What will
be the temperature of the water after it has received heat energy of
1680 kJ?

7. A cooling system contains 10 litres of water at a temperature
of 15 °C. How much heat is required to raise the water temperature
to 95 °C?

8. An engine sump contains 4 kg of oil. How much heat energy
is required to raise the oil temperature 150 °C? (Specific heat
capacity of oil = 1700 J/kg K).

9. A block of copper of mass 2·5 kg at 500 °C is plunged into a
mass of 5 kg of water at 16 °C. What is the final temperature of the
water? (Neglect heat losses and take the specific heat capacity of
copper as 400 J/kg K).

10. In motion a vehicle has a kinetic energy of 84 kJ and when
it is brought to rest, this energy is converted into heat. If all of this
heat is passed to the brake drums, calculate the temperature rise of
the drums. (Mass of drums = 8 kg, specific heat capacity = 500 J/kg K).

.2 Thermal
xpansion

When a material is heated it generally expands; this statement applies
to solids, liquids and gases. Previous studies have been concerned with
applications; now we want to determine the extent of this expansion
as it applies to solids and liquids.

Solids

Linear expansion

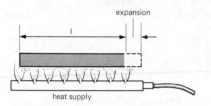

expansion

l

heat supply

Figure 2.2.1

Linear means the measurement in one dimension only, so the term linear expansion is used when the expansion in one direction only considered, e.g. the increase in length of a material when it is heate

The extent of the expansion of the rod shown in Figure 2.2.1 is governed by the type of material and the amount that it is heated. Experiments with different materials show that for general purpos the following expression may be used to find the linear expansion.

Expansion = original length x temperature rise x α

In this expression the symbol α (pronounced alpha) is called the *coefficient of linear expansion.* The value for α varies with differer materials and Table 2 gives typical values.

Table 2. Coefficients of linear expansion

Substance	Coefficient of linear expansion (per °C or per K)
Aluminium	0·000 023
Brass	0·000 019
Bronze	0·000 018
Copper	0·000 017
Steel	0·000 012
Cast iron	0·000 010
Invar steel	0·000 001

The coefficient of linear expansion is the fraction of its original length by which a material expands per degree rise in temperature

Example 1

Determine the expansion of a steel push rod of length 200 mm wh its temperature is raised by 50 °C.

$$\text{Expansion} = \text{original length x temperature rise x } \alpha$$
$$= 200 \times 50 \times 0·000\ 012$$
$$= 0·12 \text{ mm}$$

Example 2

The diameter of the skirt of a cast iron piston is 0·2 mm larger at its operating temperature than at room temperature. What would be the expansion of an aluminium piston under similar conditions

Coefficient of linear expansion: for aluminium = 0·000 023 pe for cast iron = 0·000 010 per °C. Aluminium expands 2·3 times gr than cast iron so

$$\text{expansion of aluminium piston skirt} = 0·2 \times 2·3$$
$$= 0·46 \text{ mm}$$

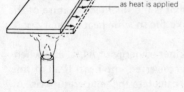

area of plate increases as heat is applied

Figure 2.2.2 Superficial expansion

Superficial expansion

When a plate (Figure 2.2.2) is heated the area increases.

Increase in area = original area x temperature rise x coefficien of superficial expansion

For general purposes

> coefficient of superficial expansion = 2 x coefficient of linear expansion

Liquids

Volumetric or cubical expansion

For calculation purposes, liquids can be treated in a similar manner to the previous examples, except that the expansion is now considered on a volume basis. So

> expansion of liquid = original volume x temperature rise
> x coefficient of cubical expansion

With liquids the value for the coefficient varies with the temperature and also depends on the container which is used. If it was possible to use a container which did not expand on heating, then the increase in volume of the liquid would appear to be greater than in the case where the container increases its size. In motor vehicle situations the container also expands, so in this event the coefficient is called the *coefficient of apparent expansion of a liquid.*

Example 3

A cooling system containing 9 litres of water at 16 °C discharges 0·324 litre of water by the overflow as the temperature is raised to 96 °C. If this discharge represents the expansion, what is the mean coefficient of cubic expansion?

Expansion = capacity (volume) x temperature rise x coefficient

$$\text{Coefficient} = \frac{\text{Expansion}}{\text{Capacity} \times \text{temperature rise}}$$

$$= \frac{0 \cdot 324}{9 \times 80} = 0 \cdot 000\ 45 \text{ per } °C$$

Coefficient of cubic expansion for this temperature range = 0·000 45/°C.

This example gives some indication of the amount that liquids expand on heating. When this is applied to such things as hydraulic brakes, it will be realised that special provision must be made to allow for the increase in volume of the liquid.

Exercises 2.2

1. A rod of length 500 mm is heated through 200 °C. What is the increase in length? (Coefficient of linear expansion = 0·000 02 per K.)

2. A rod of length 120 mm is heated through 300 °C. What is the length of the rod after heating? (Coefficient of linear expansion = 0·000 02 per K.)

3. An aluminium alloy piston operates at 250 °C whereas a piston made of cast iron normally runs at a temperature of 350 °C. Assuming both pistons are of similar construction and size, which one has the largest diameter at its operating temperature? (Coefficient of linear expansion for aluminium alloy and cast iron are $2 \cdot 3 \times 10^{-5}$ per K and $1 \cdot 0 \times 10^{-5}$ per K, respectively.)

4. An engine component has a length of 120 mm at 15 °C and has a coefficient of linear expansion of 0·000 012 per K. To what temperature is the component heated if the expansion is 0·72 mm?

43

5. A starter ring gear of diameter 300 mm at 15 °C is to be fitte
to a flywheel so as to give an interference fit of 0·6 mm. To what
temperature must the gear be heated to enable it to be fitted?
(Coefficient of linear expansion = 0·000 012 per K.)

6. A cylinder has a bore of 80 mm at 15 °C. What is the bore
when the block is heated to 95 °C? (Coefficient of linear expansion
= 0·000 01 per K.)

7. A piston has a diameter of 80 mm at 15 °C. What is the piston
diameter when it is heated to 265 °C? (Coefficient of linear
expansion = 0·000 02 per K.)

8. A piston crown has an area of 5000 mm^2 at 15 °C. If the
coefficient of superficial expansion is 0·000 04 per K, what is the
area of the crown at 265 °C?

9. An automatic transmission unit has an oil capacity of 6·4
litres (6400 cm^3) at 15 °C. What is the volume of this oil at 115 °C
assuming the coefficient of cubic expansion is 0·0004 per K?

10. The capacity at 15°C of a cooling system is 12 litres,
measured to the level of the overflow pipe. How much water will
be discharged from the system as the coolant temperature is raised
to 95 °C? (Mean coefficient of apparent cubic expansion =
0·0002 per K.)

2.3 Gas laws

Charles's Law

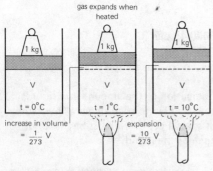

Figure 2.3.1 Expansion of gas (Charles's law)

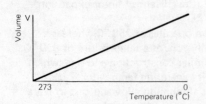

Figure 2.3.2 Graph shows volume at a given temperature — if the temperature is lowered the volume will decrease

If a gas in a cylinder is heated it will expand just like most other
substances. The ease in which a gas changes its pressure and density
is a feature which must be controlled if the expansion is to be
demonstrated.

Figure 2.3.1 shows an example of a gas being maintained at a
constant pressure by allowing the piston to slide in the cylinder;
the pressure of the gas is governed by the 'weight' acting on the
piston.

Consider the effect when heat is applied: an increase of temperature will cause the gas to expand and this will be shown by an
increase in the volume, i.e. the piston moves upwards a small
amount. By starting this experiment at 0 °C, the increase in volume
for a 1 °C rise in temperature will be 1/273 of the volume that it
occupied at 0 °C. This result is generally known as Charles's Law
which states that

**The volume of a fixed mass of gas at constant pressure expands
by 1/273 of its volume at 0°C per °C rise in temperature**

Figure 2.3.2 shows the expansion due to heating the gas and
contraction due to cooling the gas. In the latter case the volume
decreases by 1/273 for every 1 °C fall in temperature, so if this
situation continued to a temperature of −273 °C, the gas would
cease to exist. This 'gas disappearance' does not apply in practice,
because the gas changes to a liquid before the temperature of
−273 °C is reached.

A temperature of −273 °C is considered to be the temperature
at which heat ceases to exist and is called *absolute zero*. This was
proved by Lord Kelvin, and for his work in this field his name was
chosen as the SI unit for temperature. Zero on kelvin scale is

$-273\,^{\circ}$C, so to convert degrees Celsius to kelvin add 273 to the Celsius value. Temperatures based on the kelvin scale do not use the degree symbol, e.g.

$$16\,^{\circ}C = 16 + 273 = 289 \text{ K}$$

Having introduced the meaning of absolute zero, it is now possible to express Charles's Law in symbol form.
Assuming the pressure is kept constant

$$\frac{V_1}{t_1} = \frac{V_2}{t_2}$$

where
V_1 = initial volume
V_2 = final volume
t_1 = initial temperature
t_2 = final temperature

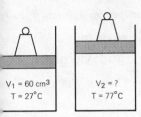

$V_1 = 60 \text{ cm}^3$
$T = 27^{\circ}C$

$V_2 = ?$
$T = 77^{\circ}C$

Figure 2.3.3

The temperature scale for t_1 and t_2 must be based on absolute zero — so the kelvin scale is used.

Example 4

Gas in the cylinder shown in Figure 2.3.3 has a volume of 60 cm^3 at 27 $^{\circ}$C. What is the volume of the gas at 77 $^{\circ}$C?

$$\text{Let } V_2 = \text{ final volume (cm}^3\text{)}$$

$$\frac{V_1}{t_1} = \frac{V_2}{t_2}$$

where
V_1 = initial volume cm^3 = 60
V_2 = final volume cm^3 = ?
t_1 = initial temperature K = 27 + 273
t_2 = final temperature K = 77 + 273

$$V_2 = \frac{V_1 t_2}{t_1}$$

$$V_2 = \frac{60 \times 350}{300} = 70 \text{ cm}^3$$

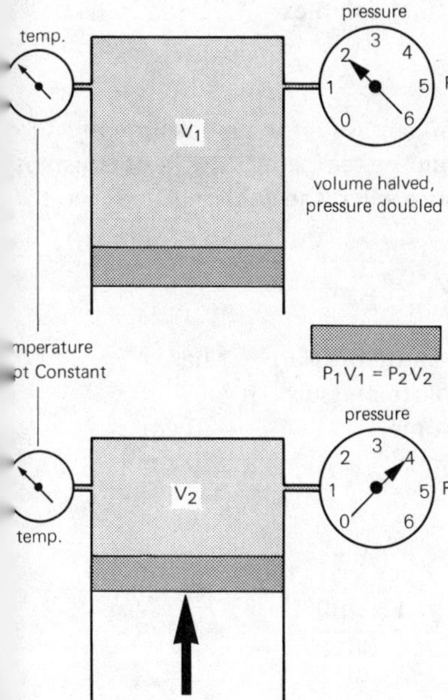

temp.

pressure

P_1

V_1

volume halved,
pressure doubled

temperature
not Constant

$P_1 V_1 = P_2 V_2$

V_2

pressure

P_2

temp.

Figure 2.3.4

Boyle's Law

Earlier studies showed that the pressure of a gas increased when the volume is decreased, assuming the temperature is kept constant. The relationship between pressure and volume is summarised by Boyle's Law which states that

The volume of a fixed mass of gas is inversely proportional to the pressure, provided the temperature remains constant

This statement can be simplified by considering Figure 2.3.4. Assuming the temperature is kept constant, the effect of decreasing the volume is to raise the pressure. Compressing the gas into half its

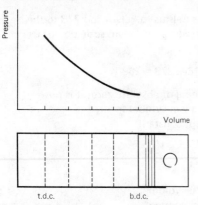

Figure 2.3.5 Graph shows that decrease in volume causes an increase in pressure

original volume doubles the pressure, and if this was continued to a point where the volume is $\frac{1}{8}$ of its original volume, i.e. a compression ratio of 8 : 1, then the final pressure would be eight times as great as the original pressure. This pressure/volume relationship is shown in Figure 2.3.5.

The pressure of the gas must be expressed as an absolute pressure. Zero on the absolute pressure scale is the point where no pressure exists, i.e. a 'perfect' vacuum. Using this scale

Standard atmospheric pressure is 101·4 kN/m^2 or 760 mm of mercury

For practical purposes this can be approximated to

atmospheric pressure = 100 kN/m^2 OR 100 kPa OR 1 bar

(The pascal, Pa, is the SI unit of pressure: 1 Pa = 1 N/m^2.)

Boyle's Law can also be expressed as

$$pV = \text{constant}$$
or
$$p_1 V_1 = p_2 V_2$$

where p_1 and p_2 = initial and final absolute pressures
V_1 and V_2 = initial and final volumes

Example 5

An engine cylinder contains 480 cm^3 of air at an absolute pressure of 1 bar (100 kN/m^2). Assuming the temperature is kept constant, what will be the absolute pressure when the volume is reduced to 60 cm^3?

$$p_1 V_1 = p_2 V_2$$

where

p_1 = initial absolute pressure = 1 bar
p_2 = final absolute pressure = ?
V_1 = initial volume = 480 cm^3
V_2 = final volume = 60 cm^3

$$p_2 = p_1 \frac{V_1}{V_2}$$

$$p_2 = \frac{1 \times 480}{60}$$

$$= 8 \text{ bar or } 800 \text{ kN/m}^2 \text{ or } 800 \text{ kPa}$$

The expression V_1/V_2 or initial-volume/final-volume is also the compression ratio of an engine. Applying this to the example (assuming the temperature remains constant) indicates that the

final pressure = initial pressure x compression ratio

During a compression test, this subject shows that a warm engine in good condition should give a minimum compression pressure of about p_1 x compression ratio.

If the throttle does not restrict the entry of air, i.e. the throttle is fully open, then the initial pressure should be about 1 bar or 100 kPa.

Since heat is generated by the compression of the gas, then the increase in pressure caused by this heat will raise the final pressure even higher.

Example 6

A compression test is conducted on an engine having a compression ratio of 9·2 : 1. Estimate the compression pressure.

Estimated initial pressure = 1 bar

Excluding effect of 'compression heat' during the test,

estimated compression pressure = 1 x 9·2

= 9·2 bar OR 920 kN/m^2 OR 920 kPa

Previous examples show that in most applications of gas laws there is an alteration to the pressure, volume and temperature of the gas. These situations can be met by combining the two gas laws to give

$$\frac{p_1 V_1}{t_1} = \frac{p_2 V_2}{t_2}$$

Example 7

Under normal operating conditions the compression stroke of a C.I. engine causes air in a cylinder of volume 1200 cm^3 to be compressed into a chamber of volume 75 cm^3. During this stroke the pressure increases from 100 kPa (1 bar) to 4000 kPa (40 bar). If the initial temperature is 87°C, what is the final temperature?

$$\frac{p_1 V_1}{t_1} = \frac{p_2 V_2}{t_2}$$

$$p_1 V_1 t_2 = p_2 V_2 t_1$$

$$t_2 = \frac{p_2 V_2 t_1}{p_1 V_1}$$

where p_1 = pressure = 1 bar
p_2 = pressure = 40 bar
V_1 = volume = 1200 cm^3
V_2 = volume = 75 cm^3
t_1 = temperature = 87 + 273 = 360 K

Substituting values gives

$$t_2 = \frac{40 \times 75 \times 360}{1 \times 1200} = 900 \text{ K} = 900 - 273 = 627 \,^\circ C$$

This example shows that if fuel oil having a self ignition temperature of about 400 $^\circ$C is injected into this heated air, then combustion of the fuel will occur.

Example 8

A tyre of a vehicle has a gauge pressure of 2 bar (200 kPa) at a temperature of 7 °C. What is the pressure of the tyre during a period of fast motoring if the tyre temperature rises to 28 °C? Gauge pressure is the pressure above atmospheric, so taking atmospheric pressure as 1 bar, then absolute pressure of tyre = 2 + 1 = 3 bars.

$$\frac{p_1 V_1}{t_1} = \frac{p_2 V_2}{t_2}$$

since volume is constant, then

$$\frac{p_1}{t_1} = \frac{p_2}{t_2}$$

∴

$$p_2 = \frac{p_1 t_2}{t_1}$$

where p_1 = initial pressure = 3 bar
p_2 = final pressure = ?
t_1 = initial temperature = 7 °C = 280 K
t_2 = final temperature = 28 °C = 301 K

$$p_2 = \frac{3 \times \overset{43}{\cancel{301}}}{\underset{40}{\cancel{280}}} = \frac{129}{40} = 3 \cdot 225 \text{ bars}$$

Final absolute pressure = 3·225 bar
Final pressure as given by a tyre pressure gauge
= 2·225 bars OR 225·5 kPa.

Exercises 2.3

1. Convert 37 °C to kelvin.
2. Convert 37 °C to degrees Celsius absolute.
3. Gas in a cylinder occupies a volume of 50 cm³ at 37 °C. If t⁻ pressure is kept constant, what is the volume of the gas at 192 °C?
4. Gas in a cylinder occupies a volume of 60 cm³ at 47 °C. If t pressure is kept constant what is the temperature of the gas when volume is increased to 180 cm³?
5. An engine cylinder contains 525 cm³ of gas at an absolute pressure of 1 bar (100 kN/m²). Assuming the temperature is kept constant, what will be the absolute pressure when the volume is reduced to 75 cm³?
6. A compression test is performed on an engine having a compression ratio of 9 : 1. What is the compression pressure, assuming none of the full charge of gas is lost and the temperature remains constant?
7. A cylinder contains a volume of 336 cm³ of gas at an absolu⁻ pressure of 100 kN/m². Assuming the temperature is kept constar what is the absolute pressure when the volume is decreased to 42 cm³? State the answer in bar and kilopascal.
8. At the start of a power stroke of an engine the gas has a vol⁻ of 50 cm³, an absolute pressure of 36 bar (3600 kPa) and a tempe⁻ ature of 1500 K. What is the temperature of the gas at the end of

the stroke when the volume is 400 cm and the pressure is 3 bar (300 kPa)?

9. A tyre has an absolute pressure of 5 bar (500 kPa) at a temperature of 27 °C. Assuming the volume remains constant, what is the pressure when the temperature is 9 °C?

10. At the start of the compression stroke the absolute pressure of the gas is 100 kN/m² and the temperature is 103 °C. What is the final temperature if the compression pressure is 1200 kN/m² absolute and the compression ratio is 8 : 1?

2.4 Changes of state

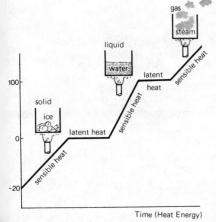

Time (Heat Energy)

Figure 2.4.1

Substances exist in one of three states; solid, liquid or gas.

A *solid* has a definite shape and volume. It resists any alteration to its shape or volume.

A *liquid* readily changes its shape to suit the container but resists any change of its volume.

A *gas* has no definite shape or volume and readily fills any container into which it is placed, but in doing so changes its density.

It is possible to alter the state of a substance by changing its temperature; the temperature at which the change occurs depends on the substance, e.g. ice changes to water at 0 °C, whereas steel does not change to a liquid until a temperature of about 1400 °C is reached.

Since water changes from solid to gas over a small temperature range this substance is normally chosen to illustrate the change of state process.

Figure 2.4.1 shows a block of ice at a temperature of −20 °C being heated at a constant rate. As heat is supplied, the temperature gradually rises, and this is shown by the graph. From −20 °C to 0 °C the heat is causing the temperature to increase so the heat is called *sensible heat,* i.e. heat which can be detected by the senses.

When the temperature reaches 0 °C, the steady rise in temperature is halted and a period of time elapses where the heat is unable to produce any increase in temperature. During this period the heat is used to give a change of state from ice to water and the heat required to produce this change is called *latent heat,* i.e. hidden heat.

Once the ice has all been melted, the temperature begins to rise again at a steady rate; the heat supplied to the water during this phase is sensible heat.

Assuming normal atmospheric conditions prevail, at 100 °C the water 'starts to boil', which is the everyday expression for the physical change of state of water to steam. The time taken, or heat energy required, for this change is far greater than that needed to change ice to water, or expressing this in a more precise manner − the latent heat of vaporization is greater than the latent heat of fusion.

As soon as the water has all become steam the temperature rise continues and the heat is once again sensible heat.

Heat energy is expressed in joules and to produce the fore-mentioned changes the following values are used

Specific heat capacity of ice   =   2·1 kJ/kg K
Specific latent heat of ice   =   336   kJ/kg K
Specific heat capacity of water =   4·2 kJ/kg K
Specific latent heat of steam   = 2260   kJ/kg K

Example 1

Calculate the heat required to raise the temperature of a mass of 2 k of water from 20 °C to 70 °C.

Heat required = mass x specific heat capacity x temperature change
= 2 x 4·2 x (70 − 20)
= 2 x 4·2 x 50
= 420 kJ

Example 2

A cooling system contains 2 litres of water. Calculate the heat required to raise the temperature from 20 °C to 100 °C and then 'boil away' 0·5 litre of the coolant.

Capacity of 2 litres = mass of 2 kg
Sensible heat   = mass x specific heat capacity
         x temperature change
     = 2 x 4·2 x (100 − 20)
     = 2 x 4·2 x 80
     = 672 kJ
Latent heat   = mass x specific latent heat of steam
     = 0·5 x 2260
     = 1130 kJ
Total heat required = sensible heat + latent heat
     = 672 + 1130
     = 1802 kJ

Physical changes which occur when substances are heated can be utilised to form temperature sensing devices, e.g. the thermostat fitted in the cooling system. Both types of thermostat, the bellows and wax capsule, operate on this principle but whereas the change of state in a bellows type is from liquid to gas, the wax capsule use the change from solid to liquid. Figure 2.4.2 shows the action of th bellows type thermostat.

The element of this type is partly filled with a liquid such as alcohol, which has a boiling point of about 80 °C under normal atmospheric pressure conditions. When the temperature reaches about 80 °C the pressure given off by the vapour causes the bellow to expand.

Earlier work showed that
1. liquids give off vapour
2. vapour pressure rises with temperature
3. boiling occurs when the external pressure equals the vapour pressure

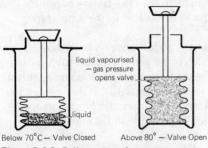

liquid vapourised
— gas pressure
opens valve

liquid

Below 70°C — Valve Closed   Above 80° — Valve Open

Figure 2.4.2 Bellows type thermostat
operation

These facts show that if a bellows type thermostat is used in a pressurized cooling system, the opening temperature will not remain constant—it will increase as the cooling system pressure increases. The wax element thermostat does not rely on the vaporization of a liquid, so the opening of the valve can be made to occur at a set temperature.

xercises 2.4

1. Calculate the heat energy required to raise the temperature of a mass of
(a) 2 kg of water from 30 °C to 50 °C
(b) 4 kg of ice from −10 °C to −5 °C
(c) 5 kg of ice at 0 °C to water at 0 °C
(d) 3 kg of water at 100 °C to steam at 100 °C
(e) 2 kg of ice at −2 °C to water at 3 °C
(f) 6 kg of water at 98 °C to steam at 100 °C

.5 Multiple choice uestions

1. The quantity of heat is the
   (a) energy in joules
   (b) hotness of a body
   (c) temperature of a substance
   (d) heat required to raise the temperature of a substance by 1 °C

2. The heat required to raise the temperature of a body by 1 °C is the
   (a) quantity of heat
   (b) mechanical equivalent of heat
   (c) temperature
   (d) heat capacity

3. The SI unit of heat capacity is the
   (a) J
   (b) K
   (c) J/K
   (d) J/kg K

4. The SI unit for specific heat capacity is the
   (a) J
   (b) K
   (c) J/K
   (d) J/kg K

5. The heat required to raise one kilogram of a substance through 1 °C is the
   (a) temperature
   (b) quantity of heat
   (c) heat capacity
   (d) specific heat capacity

6. Which one of the following values represents the specific heat capacity of water?
   (a) 900
   (b) 1700
   (c) 2100
   (d) 4200

7. A block of metal is plunged into a calorimeter containing 4 kg of water. The heat lost by the metal is
   (a) 4 x 4200 J
   (b) 16·8 kJ
   (c) mass x specific heat capacity x temperature change
   (d) mass x quantity of water x temperature increase

8. The quantity of heat liberated by burning a given
   amount of fuel is the
   (a) heat capacity
   (b) calorific value
   (c) specific heat capacity
   (d) thermal efficiency

9. Which one of the following
   is a typical calorific value
   for petrol?
   (a) 4·4 MJ/kg
   (b) 44 MJ/kg
   (c) 440 MJ/kg
   (d) 4400 MJ/kg

10. The calorific value of a fuel
    is needed for calculation of
    (a) thermal efficiency
    (b) fuel consumption
    (c) specific fuel con-
        sumption
    (d) indicated power

11. Which one of the following
    values represents the specific
    heat capacity of water?
    (a) 900 J/kg K
    (b) 1700 J/kg K
    (c) 2100 J/kg K
    (d) 4200 J/kg K

12. The heat energy required
    to raise a mass of 10 kg of
    water from 16 °C to 66 °C
    is
    (a) 500 J
    (b) 10·56 kJ
    (c) 2100 kJ
    (d) 37 632 kJ

13. Given that the specific heat
    capacity of oil is 1700
    J/kg K, the quantity of heat
    energy required to raise a
    mass of 2 kg of oil from
    15 °C to 65 °C is
    (a) 170 kJ
    (b) 1950 kJ
    (c) 33 150 kJ
    (d) 170 000 kJ

14. A material increases in
    length when it is heated.
    This expansion is called
    (a) linear
    (b) superficial
    (c) cubical
    (d) volumetric

15. The expansion of a rod is
    found by using the
    expression
    (a) final length x α
    (b) original length x α
    (c) original length x
        temperature rise x α
    (d) final length − original
        length x temperature
        rise x α

16. The fraction of a material
    original length by which a
    material expands per degr
    rise in temperature is the
    (a) expansion in length
    (b) coefficient of linear
        expansion
    (c) final length of the
        material
    (d) coefficient of appar
        expansion

17. Which one of the followir
    materials has the highest
    coefficient of linear
    expansion?
    (a) cast iron
    (b) copper
    (c) invar steel
    (d) aluminium alloy

18. The following materials h
    coefficients of linear expa
    which range between
    0·000 001 and 0·000 023
    per °K. Which material
    corresponds with the
    lower limit of this range?
    (a) cast iron
    (b) copper
    (c) invar steel
    (d) aluminium alloy

19. The expansion of some pistons is restricted by casting in a plate across the gudgeon pin bosses. This plate is made of
    (a) brass
    (b) copper
    (c) invar
    (d) bronze

20. For general purposes the coefficient of superficial expansion can be found by multiplying the coefficient of linear expansion by
    (a) 2
    (b) 3
    (c) 4
    (d) 10

21. A metal having a coefficient of linear expansion of 0·000 012 per K has a length of 200 mm at 15 °C. When it is heated to 365 °C its length will be
    (a) 0·84 mm
    (b) 0·876 mm
    (c) 200·84 mm
    (d) 200·876 mm

22. The diameter of a piston at a temperature of 265 °C is 0·3 mm larger than it was at 15 °C. If the coefficient of linear expansion is 0·000 02 per K, the piston diameter at 15 °C is
    (a) 17·67 mm
    (b) 56·6 mm
    (c) 60 mm
    (d) 166 mm

23. A gearbox has an oil capacity of 5 litres at 15 °C. What will be the increase in the volume occupied by this oil when the temperature is raised to 215 °C? (Assume a coefficient of apparent cubic expansion of 0·0004 per K.)

    (a) 0·4 cm^3
    (b) 0·43 cm^3
    (c) 400 cm^3
    (d) 430 cm^3

24. A strip of iron is placed beside a similar length of brass and both metals are heated equally. If the original length was 100 mm, what is the difference in length after a temperature rise of 300 °C? (Assume coefficients of linear expansion for brass and iron as 0·000 02 and 0·000 01 per K, respectively.)

    (a) 0·003 mm
    (b) 0·3 mm
    (c) 0·6 mm
    (d) 1·2 mm

25. A connecting rod has a length of 150 mm at 15 °C. If the coefficient of linear expansion is 0·000 012 per K the length of the rod at 215 °C is
    (a) 150·0018 mm
    (b) 150·0024 mm
    (c) 150·36 mm
    (d) 150·387 mm

26. A gas at constant pressure increases its volume by a given amount for every 1 °C rise in temperature. Raising the gas temperature from 0 °C to 1 °C causes the volume to increase by
    (a) $\dfrac{1}{100}$
    (b) $\dfrac{1}{273}$
    (c) 1 cm^3
    (d) 1 mm^3

27. Cooling a gas causes the volume to decrease. Assuming the gas did not change to a liquid, it would disappear if the temperature was lowered to
    - (a) − 460 °C
    - (b) − 273 °C
    - (c) − 100 °C
    - (d) 0 °C

28. The absolute zero of temperature is
    - (a) − 460 °C
    - (b) − 273 °C
    - (c) − 100 °C
    - (d) 0 °C

29. A temperature of 0 kelvin is
    - (a) − 460 °C
    - (b) − 273 °C
    - (c) − 100 °C
    - (d) 0 °C

30. A temperature of 43 °C is
    - (a) 503 K
    - (b) 316 K
    - (c) 143 K
    - (d) 43 K

31. The gas law $V_1/t_1 = V_2/t_2$ is known as
    - (a) Kelvin's law
    - (b) Boyle's law
    - (c) Joule's law
    - (d) Charles's law

32. Gas in a cylinder occupies a volume of 48 cm^3 at 27 °C. If the pressure is kept constant, the volume of the gas at 77 °C is
    - (a) 56 cm^3
    - (b) 137 cm^3
    - (c) 179 cm^3
    - (d) 187 cm^3

33. Gas in a cylinder occupies a volume of 50 cm^3 at 27 °C. If the pressure is kept constant, the temperature required to increase the volume to 100 cm^3 is
    - (a) 54 °C
    - (b) 150 °C
    - (c) 327 °C
    - (d) 600 °C

34. The gas law which reads: 'The volume of a fixed mass of gas is inversely proportional to the pressure, provided the temperature remains constant' is known as
    - (a) Kelvin's law
    - (b) Boyle's law
    - (c) Joule's law
    - (d) Charles's law

35. For practical purposes 'atmospheric pressure' is taken as
    - (a) 0 kN/m^2
    - (b) 15 kN/m^2
    - (c) 1 bar
    - (d) 101 bar

36. Given that atmospheric pressure is 100 kN/m^2, a gauge pressure of 150 kN/m^2 is an absolute pressure of
    - (a) 50 kN/m^2
    - (b) 250 kN/m^2
    - (c) 423 kN/m^2
    - (d) 15 000 kN/m^2

37. A tyre has an absolute pressure of 2 bar (200 kPa) at a temperature of 7 °C. Assuming the volume remains constant, what is the pressure when the temperature is raised to 28 °C?
    (a) 0·5 bar (50 kPa)
    (b) 1·8 bar (180 kPa)
    (c) 2·15 bar (215 kPa)
    (d) 8 bar (800 kPa)

38. An air storage tank has an absolute pressure of 1160 kN/m^2 at a temperature of 290 K. What is the pressure when the temperature is 300 K?
    (a) 1·2 bar (120 kPa)
    (b) 11·21 bar (1121 kPa)
    (c) 12 bar (1200 kPa)
    (d) 112 bar (11 200 kPa)

39. Given that atmospheric pressure is 100 kN/m^2, an absolute pressure of 350 kN/m^2 is a gauge pressure of
    (a) 2·5 bar (250 kPa)
    (b) 4·5 bar (450 kPa)
    (c) 250 bar (25 000 kPa)
    (d) 450 bar (45 000 kPa)

40. A pressure of 250 kPa is
    (a) 2·5 N/m^2
    (b) 250 N/m^2
    (c) 0·25 kN/m^2
    (d) 250 kN/m^2

41. A substance which resists any alteration to its definite shape and volume is a
    (a) gas
    (b) vapour
    (c) liquid
    (d) solid

42. Heat which causes a substance to increase in temperature is called
    (a) latent heat
    (b) intense heat
    (c) sensible heat
    (d) radiant heat

43. The meaning of the term 'specific latent heat of steam' is: 'The heat energy required to change a mass of 1 kg of
    (a) water to steam without change in temperature'
    (b) ice to steam without change in temperature'
    (c) steam at 100 °C to water at 99 °C'
    (d) water at 100 °C to steam at 101 °C'

44. When ice, at a temperature of −10 °C, is heated at a constant rate, the temperature rises steadily until 0 °C is reached. At this point the temperature remains constant for a period of time because the heat energy
    (a) applied is sensible heat
    (b) is producing a structural change
    (c) is changing the state from a liquid to a gas
    (d) is expanding the substance

45. As applied to ice, water, and steam, which one of the following requires the largest amount of heat energy to raise the temperature from the lower to the higher value?
    (a) − 10 °C to −8 °C
    (b) −1 °C to 1 °C
    (c) 50 °C to 52 °C
    (d) 99 °C to 101 °C

46. The specific heat capacity
of water is 4·2 kJ/kg K and
the specific latent heat of
steam is 2260 kJ/kg K.
How much heat is required
to change a mass of 2 kg
of water at 98 °C to steam
at 100 °C?
    (a)   2 kJ
    (b)   16·8 kJ
    (c)   2268·4 kJ
    (d)   4536·8 kJ

47. A bellows type thermostat
OPENS when the substance
changes its state from a
    (a)   solid to a vapour
    (b)   solid to a liquid
    (c)   liquid to a vapour
    (d)   liquid to a solid

48. A wax capsule type of
thermostat OPENS when
the substance changes its
state from a
    (a)   solid to a vapour
    (b)   solid to a liquid
    (c)   liquid to a vapour
    (d)   liquid to a solid

2.10 Combustion of petrol

Production of petrol

Petrol, or gasoline as it is sometimes called, is obtained by refining crude petroleum. Mined in various parts of the world, the crude oil, as pumped from the ground at the oil well, appears as a thick, black, dirty substance. At the refinery this dirty oil is 'cleaned' and, by means of a distillation process, the raw material is heated and divided into a number of different products. Each fuel or oil separated or fractionated in this manner has a boiling point which falls within a given range; light fractions such as petrol boil at a lower temperature than heavy products of the lubricating oil family.

The principle of distillation is shown as Figure 2.10.1. As crude oil in a flask is heated, a vapour is given off from the oil which is condensed and collected by a flask. By maintaining the oil at a temperature within the range shown in the table in Figure 2.10.1 the various products are obtained.

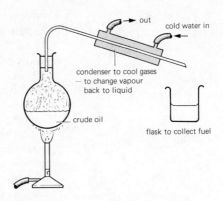

Heat supplied to crude oil	Substance given off
low ↓ high	petrol (gasoline) paraffin (kerosine) diesel oil lubricating oil bitumen

Figure 2.10.1 Distillation of a fuel

Combustion and air/fuel ratios

Petrol is a hydrocarbon, i.e. it contains carbon (C) and hydrogen (H), and when this is mixed with oxygen and ignited, a chemical change takes place which releases heat. The amount of heat liberated depends on the quantity of fuel burnt—the greater the quantity the greater is

the heat released. However, a limit is reached in an engine when all
of the oxygen in the cylinder has been consumed.

For complete combustion, 1 kg of petrol requires a mass of
15 kg of air. This value is obtained from calculations based on the
chemistry of combustion and this is generally stated as

Chemically correct air/fuel ratio for petrol is 15:1 (by mass).

When a larger proportion of air is supplied to a given quantity of
petrol, the air/fuel ratio is increased and it is said that the 'mixture
weak'. Similarly an air/fuel ratio of 12:1 is considered to be a 'rich
mixture'. If the engine uses a fuel other than petrol then the 'chem
correct' air/fuel ratio may have a value different to that stated for
petrol.

Exhaust gas composition

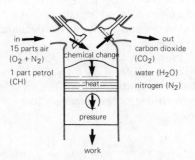

Figure 2.10.2 Inlet and outlet gas
composition

When the air/fuel ratio is correct for the fuel being used, the main
constituents of the exhaust gas are carbon dioxide (CO_2), water
(H_2O), and nitrogen (N_2). See Figure 2.10.2. If petrol is consider
to be CH, the engine causes the chemicals to change as shown in
the table.

Entering engine	Leaving engine
PETROL + AIR	EXHAUST GAS
Petrol + (Oxygen + Nitrogen)	Carbon dioxide + Water + Nitrogen
CH + O_2 + N_2	CO_2 + H_2O + N_2

A low air/fuel ratio (rich mixture) has insufficient oxygen to
complete the combustion process, so some of the fuel leaves the
cylinder in a partially burnt state. In addition to the previous
exhaust products, this rich mixture produces undesirable product
such as carbon monoxide (CO) together with hydrogen (H) and
carbon (C). It is these carbon particles emitted from the exhaust
that causes black smoke to be discharged from an engine which
is running on a rich mixture.

The exhaust product from a weak mixture is 'cleaner' than tha
given from a rich mixture. Used in this way the word 'cleaner' im
that health hazards and pollution problems are improved. It is for
this reason that carburettors should be set to operate on an air/fu
ratio that gives exhaust gas products which conform to the currer
exhaust gas emission control regulations.

The temperature attained during the combustion process is
very high and if this temperature is excessive, additional undesira
exhaust products are formed; nitrogen oxides (NO_x) is a typical
example. The emission of NO_x must be limited, so the modern
engine incorporates special features that limit combustion tempe
atures. Unfortunately this limitation on temperature normally re
in a reduction in performance of the engine.

Air/fuel ratios can be determined by an exhaust gas analyser
(Figure 2.10.3). A sample of gas is fed by a rubber hose, via a con
sation trap which extracts the steam, to an electrically operated
analyser. This instrument measures the electrical resistance of a
heated filament that is exposed to the exhaust gas. The temperat
of the filament is affected by the thermal conductivity of the gas

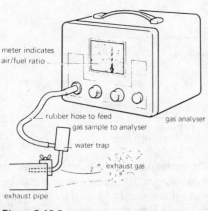

Figure 2.10.3

when the gas composition changes, the filament resistance alters and the meter indicates this effect on a scale calibrated to give the air/fuel ratio.

ffects of variations of ir/fuel ratio

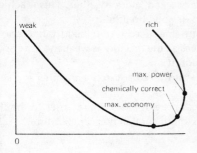

Figure 2.10.4

In an actual engine a chemically correct mixture neither produces maximum power nor gives maximum economy. This is shown by Figure 2.10.4, which illustrates the effect of varying the air/fuel ratio between the limits of very weak to very rich.

A carburettor set to give maximum economy by operating on a mixture strength which is slightly weaker than 'chemically correct' ensures that the fuel is efficiently burnt, but the slower burning rate gives lower maximum combustion temperatures and reduced power.

If the mixture is weakened beyond this point the amount of fuel required to produce one unit of power is increased and the engine may overheat due to the high heat transfer from the slow moving flame to the combustion chamber surfaces. The drop in power is clearly demonstrated when an engine is cold: pushing-in the 'choke' too early results in low engine power. In this case the very slow burning of the charge continues through the power and exhaust strokes, so when the new petrol-air mixture comes into contact with the hot exhaust gas, the incoming charge sometimes ignites and produces 'popping-back' through the carburettor.

Setting the mixture slightly richer than chemically correct gives a higher power output, but this is obtained at the expense of fuel consumption and exhaust cleanliness. When the mixture is enriched beyond this maximum power position, the fuel consumption rises a considerable amount, but the power only falls off a comparatively small amount. Figure 2.10.5 illustrates these factors.

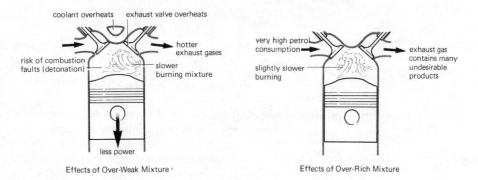

Effects of Over-Weak Mixture

Effects of Over-Rich Mixture

Figure 2.10.5

In practice the performance of an engine running on a rich mixture is often regarded as satisfactory by the driver, even although its petrol consumption has doubled, whereas a mixture which is slightly weak is soon detected. Slower burning also occurs with rich mixtures and the risk of overheating is present, but this is not so severe as that experienced with weak air/fuel ratios—the very slow burning of these weak mixtures can soon ruin an engine. For example, exhaust valves can overheat and melt ('burn'), piston seizure may occur and extensive damage can be produced by combustion faults such as pre-ignition and detonation.

2.11 Combustion faults

Normal combustion

Under normal conditions the compressed petrol-air mixture is ignit
by the sparking plug, and a flame, originating from the vicinity of
plug electrodes, progresses across the combustion chamber at a reg
rate (Figure 2.11.1). Although the combustion process is complete
in a fraction of a second, the pressure caused by the release of hea
energy rises steadily to give a smooth start to the power stroke. Th
flame speed (or flame rate) can be varied by altering the

1. compression pressure—a high flame speed is obtained when
the fuel and air particles are 'packed' close together. This is achieve
by a high compression ratio or good breathing.

2. air/fuel ratio—highest flame speed is obtained from a ratio
which is slightly rich, whereas as the mixture is weakened the spee
decreases a considerable amount.

3. degree of turbulence—air movement in the chamber increase
the flame speed.

4. quantity of exhaust gas present—acts as a retarder to the flar
and thereby lowers the maximum combustion temperature.

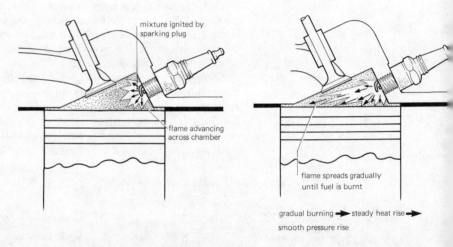

Figure 2.11.1 Normal combustion

Detonation and combustion knock

To obtain good engine power, a high flame speed is necessary, sin
the quicker the fuel burns, the higher will be the temperature of t
gas. This requirement suggests that the factors controlling flame
speed should be arranged so as to give the highest speed, but whe
this is attempted various combustion faults develop. To illustrate
this problem, let us consider an engine which has a provision for
varying the compression ratio. As the compression pressure is
increased, the flame speed is also increased, but when a certain
compression pressure is reached, the flame speed suddenly rises t
a figure which equals the speed of sound. No longer is the fuel bu
in a progressive manner; instead the petrol-air mixture explodes t
give a condition called 'detonation'. The compression pressure at
which detonation occurs depends on many factors; these include
the grade of petrol used and the type of combustion chamber.

Examination of the combustion process prior to the onset of
this severe detonation would show that a portion of the petrol-ai
mixture remote from the sparking plug (i.e. the end gas) would r

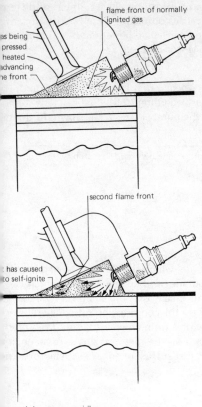

flame front of normally ignited gas

as being pressed heated advancing he front

second flame front

has caused to self-ignite

xture is burnt very rapidly.
ects — pinking, damage to engine & reduced power.
vays occurs after the spark.

ure 2.11.2 Combustion knock

be performing in the normal manner. Instead, this pocket of gas, which has been compressed and heated by the gas already burnt, will spontaneously ignite and cause a rapid build-up of pressure. This condition is called 'combustion knock' and Figure 2.11.2 indicates the main features.

Although detonation and combustion knock are two different combustion faults, the general effects are similar. For this reason the two conditions are often grouped together. In both cases they occur *after the spark.*

Effects of detonation and knock depend on the severity of the condition, but the main results are

1. pinking—a sound produced by the high pressure waves; the noise described by some people as a 'metallic tapping sound' and by others as the 'sound of fat frying in a pan'

2. shock loading of engine components

3. local overheating—piston crowns can be melted

4. reduced engine power

These effects show that detonation and knock should be avoided, and if this is to be achieved attention must be given to the items which promote detonation. The main factors are

1. compression pressure and grade of fuel—the compression ratio is linked to the fuel used; engines having high compression ratios generally require a fuel having a high octane number

2. air/fuel ratio—weak mixtures are prone to detonate

3. combustion chamber type—the degree of turbulence and provision for cooling of the end-gas have a great bearing on the compression ratio which is used

4. ignition timing—an over advanced ignition promotes detonation

5. engine temperature—overheated components increase the risk of detonation

ctane rating

The octane rating of a petrol is a measure of its resistance to knock or detonate. This anti-knock quality varies with fuels, so for classification purposes the knock resistance of a fuel is compared with two reference fuels

1. *iso-octane*—this is given the number 100 since its anti-knock quality is excellent

2. *heptane*—this has very poor resistance so the rating given is zero

A special single cylinder, variable compression engine, fitted with equipment for measuring the knock intensity, is used to determine the octane rating of a fuel. Raising the compression ratio a given amount makes the fuel knock, so by using a mixture of iso-octane and heptane it is possible to vary the proportions of these reference fuels until a similar anti-knock property is obtained. For example, if the fuel has a knock resistance equal to a mixture of 90 per cent iso-octane and 10 per cent heptane, by volume, then the octane number given to the fuel is 90. (It should be noted that the fuel being tested does not have to contain iso-octane.)

Recommendations which cover such things as induction temperature and ignition timing of the special engine ensure that the rating

given to a fuel is standardized. Unfortunately, a number of differe
methods exist and this can cause confusion, e.g. a fuel having an
octane rating of 95 based on the Research method has a rating of
84 when established by the Motor method.

Star classification

To help the British motorist select a fuel from any trade brand, B.
have classified fuels in groups according to the fuel's octane numb
Each group is identified by the number of stars as shown in the ta

Rating	Minimum octane number (Research method)
5 star	100
4 star	97
3 star	94
2 star	90

It is difficult to state the fuel which should be used in a particu
engine; each engine has an octane need or requirement which is
affected by many things. This means that the engine's 'tendency t
knock' is not only governed by the compression ratio. There is no
advantage in using a fuel having a rating higher than that required
or specified by the manufacturer—the fuel best suited to an engin
is the one which has a rating just sufficient to give satisfactory kn
free performance.

Anti-knock additives

Some fuels have excellent anti-knock qualities but in general thes
are either expensive or in short supply. Cheaper products may be
obtained by blending (mixing) these expensive fuels with other fu
which are in more plentiful supply. Alternatively, the anti-knock
quality may be improved by adding a substance such as tetra-ethy
lead (t.e.l) to the petrol. Many additives are used in a modern fue
substances containing lead are poisonous, so regulations are in for
to limit the quantity that is added.

Pre-ignition

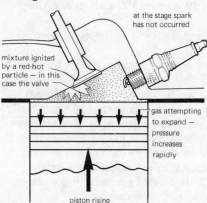

at the stage spark has not occurred

mixture ignited by a red-hot particle — in this case the valve

gas attempting to expand — pressure increases rapidly

piston rising

Figure 2.11.3 Pre-ignition (ignition before the spark)

This condition applies when the petrol-air charge in the cylinder
fired by a red-hot particle before it is ignited by the sparking plug
The incandescent object may be a carbon deposit or any protrudi
component which forms a part of the combustion chamber. Pre-
ignition always occurs *before the spark* and is an undesirable con
which can lead to detonation, melting of the piston crown and o
forms of damage. Normally pre-ignition gives a considerable redu
in the engine's power output and is often accompanied by the so
of 'pinking'.

Figure 2.11.3 shows pre-ignition of the charge, and in this cas
indicates a situation whereby the rising piston is compressing a ga
which is attempting to expand. This will result in a considerable
increase of gas pressure and combustion chamber temperature.

unning on

An engine which continues to run after the ignition is switched off is caused by the petrol-air charge being ignited by a 'hot spot', such as a sparking plug, valve or carbon deposit, which glows and fires the charge at about t.d.c. Cooling system faults and excessive carbon deposits are often responsible for this condition. (For effect of compression ratio and supercharge on engine performance and efficiency see *F. of M.V.T.*).

.12 Fuels for ompression-ignition ngines

nition delay

Whereas petrol must resist being auto-ignited by the heat generated by the compression of the gas, a fuel for a C.I. engine must ignite easily. The reason for this can be seen when the process of combustion is considered.

Towards the end of the compression stroke, the fuel is injected into the cylinder. On entering, the fuel particles extract heat from the air, vaporize and eventually self-ignite. The time taken for this phase, which is called ignition delay, may be less than a second in time, but during this period the fuel is continuing to enter the cylinder. When ignition eventually takes place, the large quantity of fuel present in the chamber is also fired. This results in a sudden pressure rise that is accompanied by the noise known as 'diesel knock'. Noise and vibration are problems which are associated with C.I. engines, so steps are taken to minimize these drawbacks; one way is to reduce the ignition delay period.

Ignition delay is affected by
1. atomization of the fuel
2. final temperature of the compressed air
3. quality of the fuel

The fuel quality needed in this context is its ability to self-ignite, i.e. the temperature at which it will spontaneously ignite. At this stage it should be pointed out that the self-ignition temperature is NOT the flash point. The *flash point* of a fuel is the lowest temperature at which the fuel gives of a vapour that will flash when exposed to a naked flame. (Flash point for diesel fuel is about 70 °C whereas the self ignition temperature is in the region of 400 °C.)

tane rating

To classify fuels for C.I. engines the *cetane* rating is adopted. This compares the ignition quality of a fuel with two reference fuels:

cetane—given the value of 100 because of its good ignition quality.
alpha-methyl-naphthalene—rated as zero.

A fuel rated at 55 has a similar ability to self-ignite to a fuel consisting of 55 per cent cetane and 45 per cent alpha-methyl-naphthalene by volume.

The fuel used in a C.I. engine should have a cetane number just high enough to give freedom from pronounced diesel knock.

.13 Engine testing

The power output of an engine and its behaviour when it is subjected to a load can be obtained when an engine is mounted to a dynamometer. Modern dynamometers load the engine either electrically or hydraulically and provide means for measuring the torque and power.

Early methods used a friction brake constructed either in the form
of two shoes (prony brake) or by a rope wound around the flywhe
(rope brake). The name given to these dynamometers show why th
power output of an engine is called the *brake power*.

Dynamometer construction

The dynamometer is an example where energy is converted from
one form to another; mechanical energy given out by the engine
is converted to
1. heat in the case of the hydraulic type
2. electrical energy when an electric type of dynamometer is u
    A number of different dynamometer arrangements are used. M
facturers and educational establishments often use test beds which
enables the engine to be tested under controlled conditions, where
in the garage an engine test can be performed in a short time by p
the vehicle's driving wheels on rollers which connect to a dynamo

Figure 2.13.1 Heenan—Froude dynamometer

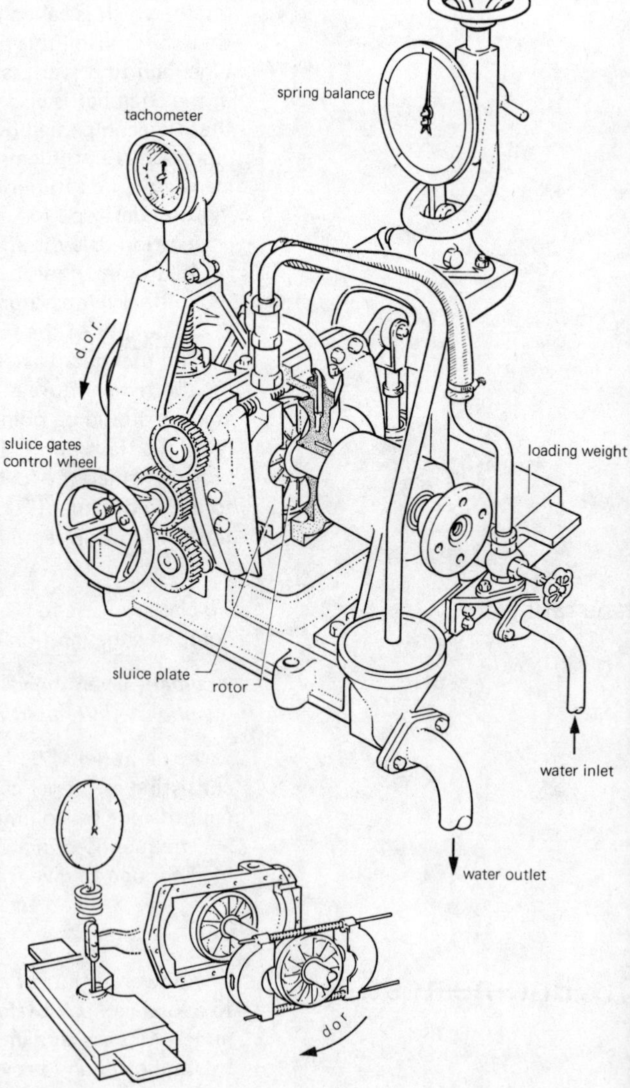

The basic construction of a Heenan-Froude hydraulic dynamometer is shown in Figure 2.13.1. A rotor, incorporating a series of radial vanes, is connected to the engine and contained in a casing, which is mounted on bearings and permitted to rotate a small amount in a similar plane to that moved by the driving shaft. Two sluice plates that are controlled by a hand wheel, expose cups, formed by the vanes of the rotor, to similar cups in the casing. The casing is supplied with water under pressure and the rate of flow is controlled by an outlet valve which should be restricted so as to maintain a pressure in the casing but opened sufficiently to prevent the outlet temperature from exceeding 60 °C.

The operation of the dynamometer is similar to that used in a fluid flywheel. Movement of the rotor causes the water to be carried around and thrown outwards into the cups in the casing. Impact with the vanes separating the cups in the casing causes the water to exert a force which tends to rotate the casing. Since the casing is resisted by a load applied at the end of the torque arm, the energy contained in the water hitting the casing is converted into heat—the greater the power developed by the engine, the greater is the heat generated.

The load applied to an engine is controlled by the sluice plates. As the plates are opened a larger area of the casing is exposed to the rotor and the load is increased.

Measurement of brake power (Figure 2.13.2) is obtained by noting the reading on the spring balance and applying this value to a formula, which is based on the following.

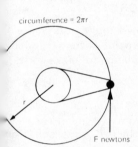

circumference = $2\pi r$

r

F newtons

gure 2.13.2

Work done by engine = force x distance moved in 1 second

if n = speed in revolutions per second

then   work = $F \times 2\pi r n$

since   torque = force x length of torque arm

i.e. $T = F \times r$

then   work = $T \times 2\pi n$ J/s

Work at the rate of 1 joule per second is equal to a power of 1 watt.

So   power = $2\pi n T$ watts

where T = torque in Nm

When a series of tests are performed on a dynamometer the expression for power can be rewritten as

$$P = Fn \times 2\pi r$$
$$P = FnK$$

where K = dynamometer constant, i.e. $2\pi r$

(Many dynamometers in use have a torque arm of length 0·356 m (14 inches) so the value of K for this size of dynamometer is 2·237.)

To summarize

$$T = F \times r$$

where T = torque (Nm)
          F = load (N)
          r = length of torque arm (m)

$$P = 2\pi nT$$

OR
$$P = FnK$$

where P = brake power (W)
  n = speed (rev/s)
  K = dynamometer constant

Example 1

An engine rotating at a speed of 3000 rev/min lifts a dynamomete
load of 87·5 N. If the length of the torque arm is 0·4 m calculate t
torque and brake power of the engine.

$$\text{Speed of 3000 rev/min} = \frac{3000}{60} = \text{rev/s}$$

$$= 50 \text{ rev/s}$$
$$\text{Torque} = \text{brake load} \times \text{length of torque arm}$$
$$= 87\cdot5 \times 0\cdot4 = 35 \text{ Nm}$$
$$\text{Brake power} = 2\pi nT$$
$$= 2 \times \frac{22}{7} \times 50 \times 35$$
$$= 11\ 000 \text{ W OR 11 kW}$$

The Imperial system uses the horsepower (hp) to rate engine
power and a prefix was used to show the point where the power
was measured; e.g. 'b.h.p.' was the abbreviation for brake horsepower.

For conversion purposes

$$1 \text{ hp} = 746 \text{ watts} \simeq 0\cdot75 \text{ kW}$$

Therefore, to convert hp to kW multiply by 0·75, for example, an
engine having a bhp of 100 has a brake power of 75 kW.

We have seen the significance of the term 'brake' as applied to
the expression 'brake power'. Using this rule it is possible to use
similar prefixes or terms to describe other locations at which the
power is considered, e.g. the total power developed by the gas
pressure in the cylinders is termed 'indicated power' (Figure 2.13
When the power that is produced in the cylinder is transmittec
to the flywheel some energy will be lost. This will be used in over
coming friction and pumping losses so

brake power = indicated power − power lost to friction and pum[

To show the extent of these mechanical losses the term mecha
efficiency is used. This is stated as

$$\text{mechanical efficiency} = \frac{\text{brake power}}{\text{indicated power}} \times \frac{100}{1} \%$$

Mechanical efficiency decreases as the engine speed is increase
a typical figure for a petrol engine is 80 per cent.

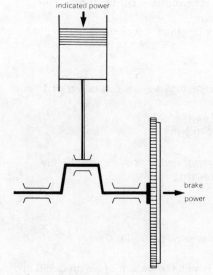

indicated power

brake
power

Figure 2.13.3

Engine performance curves

The curves shown in Figure 2.13.4 represents the power, torque a
fuel consumption of a typical petrol engine which is operating ur
maximum load conditions. To obtain these results the throttle is

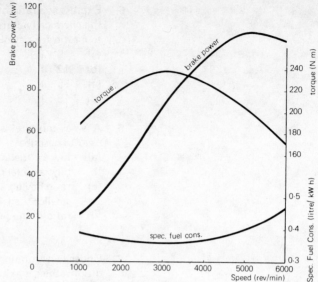

Figure 2.13.4 Power and consumption curves
(3 litre, OHC, V8 engine)

held in the full-open position and the engine speed is varied by the dynamometer load control.

Fuel consumption at each speed is measured and the specific fuel consumption is calculated. The value obtained indicates the quantity of fuel that is required to produce one unit of power, i.e.

$$\text{specific fuel consumption (s.f.c.)} = \frac{\text{fuel consumption (litres/h)}}{\text{brake power (kW)}} \text{ litres/kW h}$$

The graph shows that the speed at which the engine is using the fuel most economically is about 3000 rev/min.

Imperial system expressed s.f.c. in pints/bhp h or lb/bhp h. To convert pints/bhp h to litres/kW h multiply by 0·76, e.g.

$$0·5 \text{ pt/bhp h} = 0·38 \text{ litre/kW h.}$$

14 Multiple choice questions

1. Petrol is obtained by
   (a) refining crude oil
   (b) processing vegetable products
   (c) fractionizing coal
   (d) heating wooden substances

2. The term 'light fraction' as applied to petroleum products is a fuel which
   (a) resists vaporization
   (b) is used for lighting purposes
   (c) boils at a low temperature
   (d) boils at a high temperature

3. Petrol consists of
   (a) oxygen and carbon
   (b) carbon and hydrogen
   (c) hydrogen and nitrogen
   (d) nitrogen and oxygen

4. The minimum mass of air required to give complete combustion of 1 kg of petrol is
   (a) 9 kg
   (b) 12 kg
   (c) 15 kg
   (d) 18 kg

5. Express as a ratio of masses, the chemically correct air/ fuel ratio for petrol is
   (a) 9 : 1
   (b) 12 : 1
   (c) 15 : 1
   (d) 18 : 1

6. A 'weak mixture' is the term used to describe a
   (a) low air/fuel ratio
   (b) high air/fuel ratio
   (c) ratio having a number smaller than 15
   (d) ratio in the region of 12 : 1

7. Under ideal running con- ditions, the gas exhausted from a petrol engine using the chemically correct air/ fuel ratio consists of
   (a) carbon monoxide and nitrogen
   (b) oxygen and carbon monoxide
   (c) water, carbon monoxide and carbon dioxide
   (d) carbon dioxide, water and nitrogen

8. Which one of the following chemical symbols represents carbon monoxide?
   (a) CNO
   (b) NO_x
   (c) CO
   (d) CO_2

9. Which one of the following is an exhaust emission which is associated with high combustion temperatures?
   (a) CO
   (b) CO_2
   (c) NO_x
   (d) H_2O

10. The majority of electrical operated exhaust gas analysers operate by measuring the
    (a) thermal conductivity of the gas
    (b) temperature of the
    (c) speed of the gas
    (d) mass of the gas

11. To obtain maximum power from an actual engine the air/fuel ratio should be set
    (a) slightly richer than the chemically correct value
    (b) slightly weaker than the chemically correct value
    (c) to the chemically correct value
    (d) to give a weak mixture at low speed and rich mixture at high speed

12. When does 'combustion knock' and 'detonation' occur?
    (a) Before the spark
    (b) Just before the spark
    (c) At the time of the spark
    (d) After the spark

13. Which one of the following is associated with 'combustion knock'?
    (a) Pinking
    (b) Missing
    (c) Popping back
    (d) Explosions in the exhaust

14. Which one of the following fuels would cause a high compression engine to knock? A fuel which has
    (a) low octane value
    (b) high octane value
    (c) low calorific value
    (d) high calorific value

15. An octane rating of 80 means that the fuel
    (a) boils at a temperature of 80 °C
    (b) has a flash point of 80 °C
    (c) is made up of 80 per cent iso-octane and 20 per cent heptane
    (d) has a similar anti-knock property to a reference fuel consisting of 80 per cent iso-octane by volume

16. A fuel rated by BSI as 5 star has an octane rating of
    (a) 90
    (b) 94
    (c) 97
    (d) 100

17. Which one of the following is classed as an anti-knock additive to petrol?
    (a) Tetra-ethyl-lead
    (b) Alpha-methyl-naphthalene
    (c) Iso-octane
    (d) Cetane

18. When the petrol-air charge in an engine cylinder is fired by a red-hot particle the condition is called
    (a) knock
    (b) detonation
    (c) pre-ignition
    (d) hunting

19. The period of time between injection and initial burning of the fuel in a C.I. engine is called
    (a) flame spread
    (b) ignition delay
    (c) diesel knock
    (d) flash time

20. As applied to a fuel, the term 'self-ignition temperature' means the

(a) flash point
(b) cetane rating
(c) temperature at which the fuel gives off an inflammable vapour
(d) temperature at which the fuel will spontaneously ignite

21. A typical self-ignition temperature for a fuel used in a C.I. engine is
    (a) 70 °C
    (b) 180 °C
    (c) 400 °C
    (d) 552 °C

22. The ignition quality of a fuel for a C.I. engine is called the
    (a) delay period
    (b) knock value
    (c) cetane rating
    (d) octane rating

23. The sudden pressure rise resulting from a long delay period during the combustion process in a C.I. engine causes
    (a) diesel knock
    (b) rapid atomization of the fuel
    (c) ignition of the fuel
    (d) pre-ignition

24. Which one of the following reduces the ignition delay in a C.I. engine?
    (a) Lowering the temperature of the air charge
    (b) Reducing the atomization of the fuel
    (c) Using a fuel with a higher cetane value
    (d) Using a fuel with a higher self-ignition temperature

25. Which one of the following reduces 'diesel knock'?
    (a) Lowering the temperature of the air charge
    (b) Reducing the atomization of the fuel
    (c) Using a fuel with a higher cetane value
    (d) Using a fuel with a high self-ignition temperature

26. A fuel has a cetane value of 55. This means that the fuel
    (a) contains 55 per cent of cetane by volume
    (b) contains 55 per cent cetane by mass
    (c) has a flash point similar to a fuel consisting of 55 per cent cetane and 45 per cent alpha-methyl-naphthalene by volume
    (d) has a self-ignition temperature similar to a fuel consisting of 55 per cent cetane and 45 per cent alpha-methyl-naphthalene by volume

27. A dynamometer is an instrument for measuring the
    (a) energy output of a machine
    (b) energy given out by an electrical machine
    (c) effort required to drive a machine
    (d) water flow through an engine

28. The Heenan-Froude hydraulic dynamometer converts mechanical energy to
    (a) heat energy
    (b) electrical energy
    (c) drive a prony brake
    (d) drive a rope brake

29. The load applied to an engine by a Heenan-Froude hydraulic dynamometer is controlled by the
    (a) torque arm
    (b) sluice plates
    (c) casing position
    (d) spring balance setting

30. The power output from an engine is called the
    (a) torque
    (b) dynamometer load
    (c) indicated power
    (d) brake power

31. If T = torque (Nm) and n = speed (rev/s), which one of the following expressions represents the power in watts?
    (a) $\pi n T$
    (b) $2\pi T$
    (c) $2n T$
    (d) $2\pi n T$

32. If the torque output of an engine is known, the brake power can be calculated by multiplying the torque in newton-metres by
    (a) $2\pi r$, where r = radius in metres
    (b) $2\pi n$, where n = speed in revolutions per second
    (c) the speed
    (d) the torque arm length

33. A load of F newtons is lifted by a dynamometer having a torque arm length of r metres. The torque acting on the dynamometer is
    (a) $F \times r$ joules
    (b) F/r joules
    (c) $F \times r$ newton metres
    (d) F/r newton metres

34. Given the following engine dynamometer readings calculate the brake power in kilowatts: load lifted = 84 N, length of torque arm = 0·5 m, speed = 100 rev/s. (Take π as 22/7)
    (a)   13·2
    (b)   26·4
    (c)   4200
    (d)   8400

35. An engine develops a torque of 84 Nm at a speed of 3600 rev/min. Taking π as 22/7, the brake power is
    (a)   1·9008 kW
    (b)   15·84 kW
    (c)   31·68 kW
    (d)   302·4 kW

36. An engine lifts a dynamometer load of 120 N. If the torque arm length is 0·4 m, the engine torque is
    (a)   3·3 Nm
    (b)   48   Nm
    (c)   300 Nm
    (d)   301·7 Nm

37. The brake power of an engine is 88 kW at a speed of 70 rev/s. Taking π as 22/7 the engine torque is
    (a)   0·2 Nm
    (b)   200 Nm
    (c)   400 Nm
    (d)   1257·1 Nm

38. An engine tested against a dynamometer having a constant of 2·237, lifts a load of 100 N at a speed of 80 rev/s. If the power is given by $P = FnK$, the brake power of this engine is
    (a)   1·125 kW
    (b)   17·896 kW
    (c)   56·245 kW
    (d)   112·489 kW

39. An engine has a brake power of 60 kW and an efficiency of 75 per cent. The indicated power is
    (a)   0·8 kW
    (b)   45 kW
    (c)   80 kW
    (d)   125 kW

40. The term 'brake' as applied to the expression 'brake power' is used to show
    (a)   the type of movement given by the engine
    (b)   the point at which the power is measured
    (c)   that a retarding action is applied to the flywheel
    (d)   that the energy is used to act on a brake

2 SCIENCE (c) Mechanics and machines

2.20 Friction and lubrication

Friction

Static and sliding friction

When one surface rubs against another surface the resistance to sliding is called friction. It is possible to investigate friction by using the experiment shown in Figure 2.20.2.

A block is placed on a dry horizontal table and a force is applied to slide the block along. When the force is sufficient to overcome the interlocking action of the contacting surfaces, the block moves. As soon as it starts to move, the force required to keep it moving is less than that required to 'unstick it'. This shows that

Sliding friction is less than static friction

M.V. applications of static and sliding friction are as follows.
1. Braking—A wheel skidding over a road surface produces a smaller braking action than a wheel held on the verge of skidding.
2. Crankshaft—The torque required to initially turn a crankshaft is greater than that required to keep it moving. This fact shows up when new bearings and pistons have been fitted. (Sometimes the expression 'stiction' is used to describe the effect of static friction.

Coefficient of friction

Investigation of the forces required to slide the block Figure 2.20. along a horizontal surface reveals that a relationship exists between the force *F* needed to overcome friction and the force *W* pressing the surfaces together. The constant *F/W* is called the *coefficient of friction* and is represented by the Greek letter μ

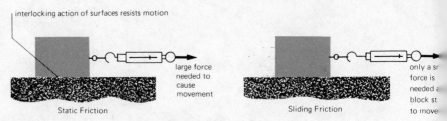

interlocking action of surfaces resists motion

large force needed to cause movement

Static Friction

Sliding Friction

only a small force is needed a block st to move

Figure 2.20.1 Sliding friction is less than static friction

$$\text{Coefficient of friction} = \frac{\text{sliding frictional force}}{\text{force between surfaces}}$$

or

$$\mu = \frac{F}{W}$$

or

$$F = \mu W$$

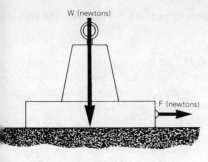

W (newtons)	F (newtons)	$\frac{F}{W}$
4	2	$^2/_4$ = 0.5
6	3	$^3/_6$ = 0.5
8	4	$^4/_8$ = 0.5
10	5	$^5/_{10}$ = 0.5
12	6	$^6/_{12}$ = 0.5

gure 2.20.2 Relationship between F and
W (F/W is equal to the
coefficient of friction)

In Figure 2.20.2 the value for μ is 0·5 and since it is a ratio no units are used.

The coefficient of friction varies with different materials; Table 3 shows some typical values.

Table 3

Surfaces	μ
Tyre on normal road surface	0·6
Brake lining on cast iron drum	0·4
Clutch lining on cast iron flywheel	0·35
Metal on metal (dry)	0·2
Metal on metal (lubricated)	0·1

In some situations friction is useful but in many other cases it must be reduced by one or more of the following

1. employing low-friction materials
2. lubrication
3. using ball or roller bearings since rolling friction is far less than sliding friction

aws of friction

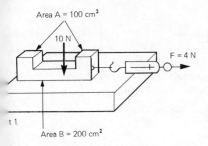

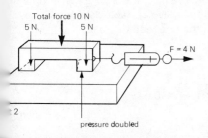

ure 2.20.3 Friction is independent of
the area in contact

Simple experiments show that friction between surfaces obeys the following laws

1. The frictional force is proportional to the force pressing the surfaces together.
2. It depends on the nature of the surfaces.
3. It is independent of the area in contact.
4. It is independent of the speed of rubbing.

The nature of the surface means the type of material and condition of the surfaces, i.e. wet, dry, smooth or rough.

The simple tests (Figure 2.20.3) show that the area in contact does not affect the frictional force. In Test 1 the spring balance may be considered as pulling along four blocks; each one carrying a load of 2·5 N, whereas in Test 2 the balance is pulling two blocks each having a load of 5 N.

On a motor vehicle the friction law referring to area has to be applied carefully, since other factors often affect the result. These factors apply to

1. Brakes—increasing the lining width of a brake shoe does not increase the friction. However, the larger area gives a reduction in the rate of wear and also allows the lining to operate at a lower temperature — this is beneficial since the coefficient of friction decreases when a given temperature is reached.

2. Bearings—increasing the diameter or length of a plain bearing does not increase the friction, that is, assuming that the bearing is operated in a dry condition. Normally bearings are lubricated, so the basic laws of friction do not apply. Instead 'fluid friction' laws apply, and with these laws an increase in the bearing area produces a larger 'oil drag'.

3. Tyres—a smooth tyre running on a smooth dry road has the same friction value as a similar tyre of wider section. Once again this is a situation which should not be found in practice. A tyre has a tread which cuts through the greasy, wet film on the surface and

'bites' into the road. So if the area of tread is increased, the grip is generally improved.

Effect of temperature on friction

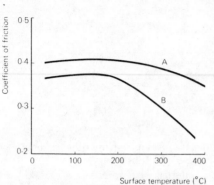

Figure 2.20.4 Effect of temperature on brake linings (lining A fades at a higher temperature than does B)

Most modern asbestos materials used as brake linings and clutch facings maintain a near-constant friction value for operating temperatures up to about 300 °C. When this temperature is exceeded, the coefficient of friction decreases and 'fade' is said to occur. As applied to brakes, fade results in an increase in the pedal force, and in cases where the brakes have become overheated, the driver experiences difficulty in stopping the vehicle.

Figure 2.20.4 shows the effect of temperature on two different asbestos based materials. Tests are performed by the manufacturer to select a material which suits the particular vehicle, so if an inferior cheap substitute is subsequently used, fade may occur during normal brake operation.

Example 1

A brake pad is pressed against a disc by a force of 400 N. If the coefficient of friction is 0·4 and the pad acts at a radius of 150 mm (Figure 2.20.5), calculate the

(a) frictional force
(b) braking torque given by the pad

$$\mu = \frac{F}{W} = \frac{\text{frictional force}}{\text{force between surfaces}}$$

$$F = \mu W$$

$$F = 0·4 \times 400$$

∴ frictional force = 160 N

$$\begin{aligned} \text{Torque} &= \text{force} \times \text{radius} \\ &= 160 \times 0·15 \quad (\text{since } 150 \text{ mm} = 0·15 \text{ m}) \\ &= 24 \text{ Nm} \end{aligned}$$

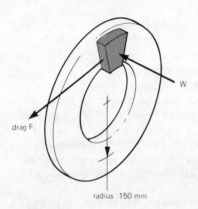

radius 150 mm

Figure 2.20.5

Example 2

When all wheels are locked on a vehicle of weight 8 kN, a force of 5 kN is required to drag it along a level surface. Calculate the coefficient of friction between the tyres and the road.

$$\mu = \frac{F}{W}$$

$$\mu = \frac{5}{8} = 0·625$$

Lubrication

Examination of two metals that have rubbed together for a period of time shows that

1. heat is generated — this indicates that energy has been used overcoming friction

2. wear or scoring has occurred due to the high spots on the two surfaces overheating and consequently fusing together

The extent of these effects is governed by the friction, so if energy losses and wear are to be kept to a minimum, provision must be made to reduce the friction. One way is to lubricate the surfaces, and no doubt you can give many instances where a spot of oil from an oil-can has eased the working of some mechanism — the lubricant in these cases will have made the surfaces smooth or slippery and this, in turn, will have reduced the friction.

Low friction is achieved when the surfaces are held apart. A fluid which does this is called a lubricant. Water on a road surface acts as a lubricant to a rubber tyre and air is used in some bearings of industrial machinery, but the most common lubricant used on motor vehicles is oil. This substance is obtained from mineral deposits, animal fats, or vegetable products; the three types are not mixable and each has particular applications. (See *F. of M.V.T.* — Lubrication)

Properties of oil

Two important properties of oil are oiliness and viscosity.

Oiliness. A spot of oil applied to a clean piece of metal soon runs into the small crevices in the surface and this results in a 'clinging action' or attraction of an oil to the metal surface. This action is called 'oiliness' and the property varies with oils, e.g. vegetable based oils have excellent oiliness qualities. A method which relies on the 'oiliness' to provide an oil film is termed 'boundary lubrication' (Figure 2.20.6).

Viscosity. Oil particles tend to cling to each other and this produces a resistance to flow which is called 'viscosity'. This property can be measured in many ways, and Figure 2.20.7 shows

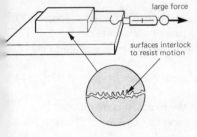

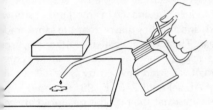

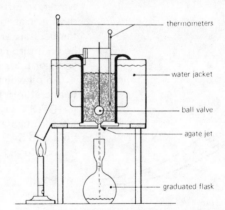

Figure 2.20.7 Measurement of oil viscosity by the Redwood viscometer

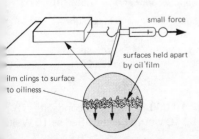

Figure 2.20.6 Boundary lubrication

the principle of one viscometer, the Redwood, commonly used in Great Britain. The viscosity of the oil, expressed in Redwood seconds, is the time taken for 50 cm^3 to flow through an orifice of area 2 mm^2. Temperature affects the viscosity, so the test should be performed at specified temperatures.

The rating introduced by the American Society of Automotive Engineers (SAE) is universally adopted to classify oils according to their viscosities. This method is based on a test performed with a Saybolt viscometer at a temperature of 99°C (210°F) Table 4 shows approximate Redwood values.

TABLE 4. SAE Crankcase Oil Classification

SAE No.	Viscosity range (Redwood seconds at 200 °F (93 °C))	
	Minimum	Maximum
20	43	55
30	55	67
40	67	83
50	83	112

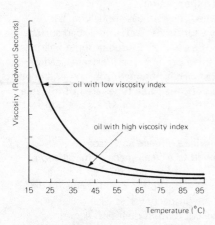

Figure 2.20.8 Change in viscosity with temperature

An engine oil having an SAE number of 50 has a higher viscosit̲ or is 'thicker' than a SAE 20 oil.

Viscosity Index. An oil gets 'thinner' as the temperature increases, but this decrease in viscosity varies with different oils (Figure 2.20.8). To indicate the effect of temperature on the viscosity of an oil, an index based on the behaviour of two oils is used:

(a) Pennsylvanian oil — small viscosity variation — index 100

(b) Gulf Coast oil — large viscosity variation — index 0

By comparing other oils with these two it is possible to give eac̲ oil an index.

Modern engine oils have a viscosity index of at least 80 and in some cases greater than 100.

Multi-grade Oils. In the past the high viscosity oil used in an engine during the summer made engine cranking difficult in the winter, so different grades were specified for the two seasons. Now̲ days special additives that reduce the change in an oil's viscosity w̲ temperature are often used, and this has meant that the same grade̲ be used throughout the year. These oils are called multi-grade, cro̲ grade or trade names which *suggest* that the viscosity remains cons̲

An oil which has this small variation in viscosity is tested at two̲ temperatures; the original SAE value of 210 °F (99 °C) and 0 °F (−18 °C). The letter W indicates that the rating is measured at the sub-zero temperature. Figure 2.20.9 shows how a multi-grade oil i̲ similar to SAE 20 at a low temperature and SAE 50 at a high temperature.

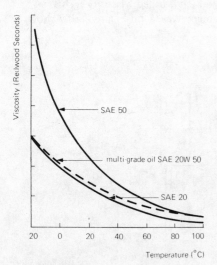

Figure 2.20.9 Effect of temperature on multigrade oil

TABLE 5. SAE Crankcase Oil Classification

SAE No.	Viscosity range (Redwood seconds at 0 °F (−18 °C))	
	Minimum	Maximum
5 W	−	3520
10 W	5250	10 560
20 W	10 560	42 000

An oil having a viscosity of 10 000 Redwood seconds at −18 ° and 70 Redwood seconds at 93 °C would be classified as 10 W 40

Transmission oils

Gear oils are subjected to conditions different from that of the engine, so they are classified under a different specification.

Resistance to wear under severe loading and other necessary

qualities of an oil are improved by using special additives.
(See *F. of M. V. T.* — Additives and Spiral bevel lubrication)

Grease

Grease is used when it is difficult to retain a less viscous lubricant.
The grease is usually manufactured by mixing a 'heavy' oil to a
substance called a base, which is selected to suit the operating
conditions. The following bases are commonly used: limesoap for
general work, bentone for hubs which are subjected to high temperature and lithium for resistance to corrosion.

Some grease bases do not mix, so it is advisable to follow the
manufacturer's recommendations.

2.21 Friction clutches

The widespread use of a friction clutch to perform a number of tasks
on a motor vehicle means that the basic principle of operation should
be understood if accurate fault diagnosis is to be carried out.

One essential feature of a transmission clutch is that is should
transmit the required torque without slipping. Whenever slip occurs,
energy that would have been available for useful purposes will now be
lost. An indication of the extent of this energy loss is shown up by the
amount of heat generated at the clutch.

Torque transmitted by plate type clutch

Figure 2.21.1 shows a pad which is pressed against a disc by a force
p newtons. When a spring balance is used to drag the pad across the
disc, it is found that the reading increases up to the point where the
pad slides. This occurs when

$$\text{frictional force } F = \mu p$$

The frictional force F may be considered as the force which acts
at the centre of the block. This force is at a radius of r, so the torque
required to give slip is given by

$$\text{torque} = \text{force x radius}$$
$$T = F \times r$$
$$T = \mu p \times r$$
$$T = \mu p r$$

Replacing the pad with an annular ring (Figure 2.21.2) having the
same width as the pad will not alter the torque, because friction is
independent of the area in contact.

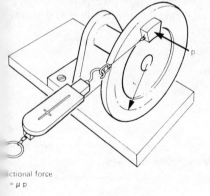

frictional force
$= \mu p$

Figure 2.21.1

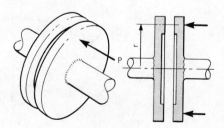

Figure 2.21.2

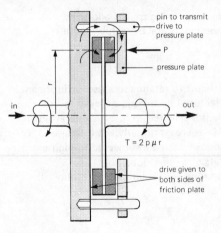

Figure 2.21.3

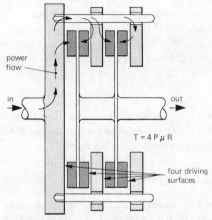

Figure 2.21.4

Sandwiching a friction disc between two driving surfaces (Figure 2.21.3) will double the torque required to produce slip, since this arrangement utilises the friction on both sides of the disc.

If the construction shown as Figure 2.21.4 was used, the torque that could be transmitted before slip developed would be four times the original value.

To summarise

$$\text{torque transmitted} = \mu pr \times \text{number of contacts}$$

Symbols make this relationship easier to remember

$$T = Sp\mu r$$

where T = torque transmitted (N m)
S = number of contacts
p = total spring thrust (N)
μ = coefficient of friction
r = mean radius (m)

The power transmitted by a clutch is given by

$$\text{power} = \text{work done by the clutch in one second}$$
$$= T \times 2\pi n$$

or $$P = 2\pi nT$$

where P = power (watts)
n = speed (rev/s)
T = torque (N m)

Example 1

Calculate the torque transmitted by a single plate clutch having a mean radius of 100 mm, total spring thrust of 1500 N and coefficient of friction of 0·3

$$T = Sp\mu r$$
$$T = 2 \times 1500 \times 0.3 \times 0.1$$

(a single plate clutch has two driving surfaces)

$$T = 90 \text{ N m}$$

Example 2

What is the effect on the torque transmitted when the lining A in Figure 2.21.5 is replaced by the wider lining B?

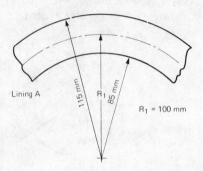
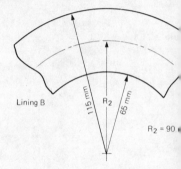

Figure 2.21.5

The mean radius of

$$\text{lining A} = \frac{115 + 85}{2} = 100 \text{ mm}$$

$$\text{lining B} = \frac{115 + 65}{2} = 90 \text{ mm}$$

Since lining B has a mean radius which is 90 per cent of lining A, the torque transmitted by a clutch fitted with the wider lining is lower to the extent of 10 per cent.

These examples can be applied to practical situations. Example 1 shows that if slip is present the fault must be due to a reduction in one or more of the factors: S, p, μ, or r. Inspection of these reveals that p and μ are the likely culprits — anything which reduces the total spring thrust or the coefficient of friction can be the cause of clutch slip.

The example of the lining width is intended to dispel a belief that an increase in area allows more torque to be transmitted. A suitable width of lining is one which is sufficiently narrow to give the largest mean radius but not so narrow as to allow rapid wear or fade to occur.

To prevent slip occurring at the point of maximum engine torque, the clutch must be capable of handling more torque than that produced by the engine. In general this safety factor is 25 per cent for private cars and 40 per cent in the case of vehicles used for commercial purposes.

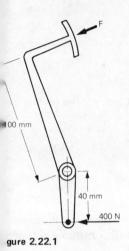

00 mm

40 mm

400 N

gure 2.22.1

2.22 Moments and centre of gravity

oments

Earlier work has shown how the principle of moments can be applied to many motor vehicle situations. The principle of moments states that when a body is in equilibrium, the clockwise moments about any point equals the anti-clockwise moments about the same point.

Example 1

Calculate the force required on the pedal shown in Figure 2.22.1.

$$\text{Anti-clockwise moment} = 400 \times 40 \text{ N mm}$$
$$\text{clockwise moment} = F \times 100 \text{ N mm}$$
$$\text{clockwise moment} = \text{anti-clockwise moment}$$
$$F \times 100 = 400 \times 40$$
$$F = \frac{400 \times 40}{100} = 160 \text{ N}$$

Example 2

A brake pedal acts on a linkage as shown in Figure 2.22.2 with pivots at P_1, P_2 and P_3. Determine the effort at the pedal to produce a force of 600 N on rod A (friction may be neglected).

Taking moments about P_3 to find force F_b on rod B

$$F_b \times 100 = 600 \times 50$$
$$F_b = \frac{600 \times 50}{100} = 300 \text{ N}$$

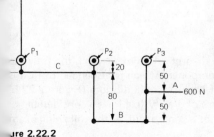

P_1 C P_2 20 P_3 50 A 600 N 80 50 B

ure 2.22.2

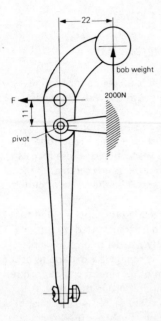

Figure 2.22.3

Taking moments about P_2 to find force F_c in rod C

$$F_c \times 20 = 300 \times 100$$

$$F_c = \frac{300 \times 100}{20} = 1500 \text{ N}$$

Taking moments about P_1 to find force F on pedal pad

$$F \times 100 = 1500 \times 20$$

$$F = \frac{1500 \times 20}{100}$$

$$F = 300 \text{ N}$$

Example 3

Figure 2.22.3 shows a clutch release lever. Calculate the force F.
Taking moments about the pivot

$$F \times 11 = 2000 \times 22$$

$$F = \frac{2000 \times 22}{11}$$

$$F = 4000 \text{ N}$$

In examples of this type it must be remembered that the momer
is the product of the force and the *perpendicular* distance between
the line of action of the force and the pivot.

Couples

When two parallel forces act in opposite directions a *couple* is form
Figure 2.22.4 gives two examples of couples.

The magnitude of a couple is

force x perpendicular distance between the forces

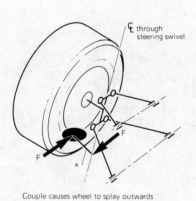

Couple causes wheel to splay outwards

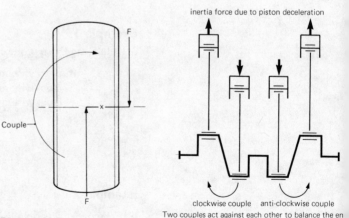

Figure 2.22.4

Centre of gravity

The centre of gravity is the theoretical point at which the total ma
is considered to act. Any component which is suspended from this
point would be in a state of balance or equilibrium (Figure 2.22.5)

The position of the centre of gravity (c.g.) has a considerable
influence on the behaviour of a component, as the following exam
show.

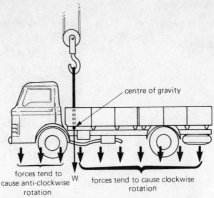

Figure 2.22.5 Vehicle will be in a state of balance if suspended from the centre of gravity

Vehicle. The position of the centre of gravity of a vehicle is controlled by the location of its main components and occupants. The distances between the c.g. and the front and rear wheels control the load carried by the axles. In cases where the c.g. is towards the front of the vehicle, for example, in front wheel drive vehicles, the load taken by the rear wheels is small. This will influence such things as tyre inflation pressures and the degree of braking which can be provided by the rear wheels (Figure 2.22.6).

Height of the c.g. affects the angle of roll when a vehicle is cornering (Figure 2.22.7).

Figure 2.22.6

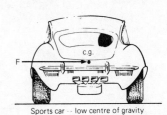

Vehicle loaded in a manner which gives high centre of gravity

Sports car -- low centre of gravity

Figure 2.22.7

Rotating components. Any part of a vehicle which revolves at speed (shafts, wheels, etc.) should have its c.g. on the axis of rotation (Figure 2.22.8). If this condition is not achieved, an out-of-balance force is produced which causes considerable vibration.

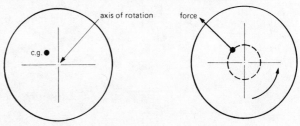

Figure 2.22.8 An out-of-balance force causes vibration if the c.g. does not coincide with the axis of rotation

Determination of position of centre of gravity

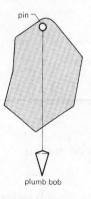

pin

plumb bob

centre of
gravity at point
where lines cross

Figure 2.22.9 Determination of position of
centre of gravity by suspen-
sion method

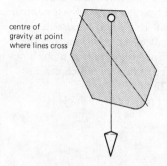

Location of the c.g. of an object having a uniform thickness can b
determined by the method shown in Figure 2.22.9. This method
would be difficult to apply to a complete vehicle so other method
are used as shown by the following examples.

Example 1

The front wheels of a vehicle of mass 300 kg support a load of 2 k
(Figure 2.22.10). If the wheelbase is 2·7 m, calculate the position
of the c.g. (Take g as 10 m/s^2.)

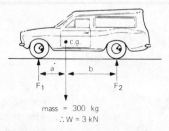

mass = 300 kg
∴ W = 3 kN

Figure 2.22.10

Taking moments about the rear wheel contact point

$$\text{Anticlockwise moment} = W \times b$$
$$\text{clockwise moment} = F_1 \times (a + b)$$
$$W \times b = F_1 \times (a + b)$$
$$\therefore \quad b = \frac{F_1 \times (a + b)}{W}$$

Substituting values

$$b = \frac{2 \times 270}{3} = 180 \text{ cm OR } 1·8 \text{ m}$$

$$\therefore \quad a = \text{wheelbase} - b$$
$$= 2·7 - 1·8 = 0·9 \text{ m}$$

The c.g. is situated 0·9 m behind the front wheels and 1·8 m
in front of the rear wheels.

Example 2

Suspending a connecting rod by two spring balances gave results a
shown in Figure 2.22.11. Calculate the position of the c.g. from th
centre of the big-end.

$$\text{Total weight of connecting rod} = F_1 + F_2$$
$$= 4 + 12 = 16 \text{ N}$$

Taking moments about the centre of the big-end

$$W \times b = F_1 \times (a + b)$$
$$\therefore \quad b = \frac{F_1 \times (a + b)}{W}$$

Substituting values

$$b = \frac{\overset{50}{\cancel{4} \times \cancel{200}}}{\underset{4}{\cancel{16}}} = 50 \text{ mm}$$

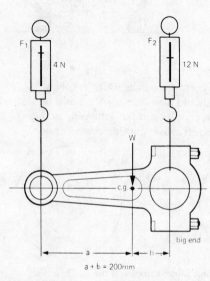

F_1 4 N F_2 12 N

W

c.g.

big end

a + b = 200mm

Figure 2.22.11

Example 3

A wheel is supported by a pivot which is located at the wheel's axis of rotation (Figure 2.22.12). In this position an out-of-balance mass of 50 g acts at a radius of 327 mm. What mass must be placed on the rim at a radius of 150 mm to make the c.g. coincide with the wheel axis?

When balance is correct (i.e., c.g. is on wheel axis)

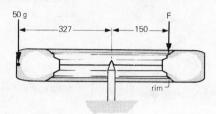

Figure 2.22.12

anticlockwise moments about pivot = clockwise moments about pivot

$$\therefore \qquad 50 \times 327 = F \times 150$$
$$F = \frac{50 \times 327}{150}$$
$$F = 109 \text{ g}$$

A mass of 109 g placed on the rim diametrically opposite the out-of-balance mass will give 'static balance'.

.23 Centrifugal ɔrce and balancing

ɛntrifugal force

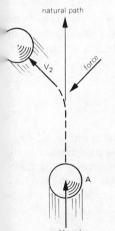

Figure 2.23.1

Newton's first law of motion states that a body remains in a state of rest or of uniform motion in a straight line unless it is acted on by some external resultant force.

This statement shows that the natural path taken by a moving object is a straight line. To make the object depart from this path, a force must be applied. Let us consider the practical effects of this force.

Figure 2.23.1 shows a steel ball in position A moving at v_1 m/s. To change the direction of motion, a force is applied to the ball and this gives a movement as shown in position B.

If the ball is attached to a cord and rotated about a point (Figure 2.23.2), then the direction of motion will change continually. The diagram shows the motion of the ball at positions A and B, and from this it will be seen that an inward acting force must be applied by the cord if the ball is to move in a circular path. A force that acts towards the centre of rotation is called 'centripetal force'.

Newton's third law of motion states that to every acting force there is always an equal and opposite reactive force.

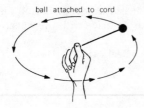

ball attached to cord

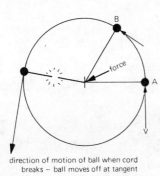

direction of motion of ball when cord
breaks – ball moves off at tangent

Figure 2.23.2

When this law is applied to an object moving in a circular path, the reactive force is caused by the inertia of the body and acts in an outward direction. The force is called 'centrifugal force' and exists whenever a centripetal force causes an object to move in a circular path.

Breakage of the cord in Figure 2.23.2 will cause the ball to 'fly-off' at a tangent to the circle: it moves off on its natural path, a straight line, because centripetal force is absent.

When the relationship between centripetal and centrifugal force appreciated, the reader will discover many cases where the term 'centrifugal' is used loosely to describe the outward movement of a rotating item, e.g. governors, ignition timing controls, fluid coupling etc.

Centrifugal force is given by the formula

$$F_c = \frac{mv^2}{r}$$

where F_c = centrifugal force (N)
  m = mass (kg)
  v = linear velocity (m/s)
  r = radius (m)

In most applications the speed of rotation is given and, to include this, the formula may be converted as follows

let v = linear velocity (i.e. velocity in a straight line)
if n = rotational speed (rev/s)
then v = circumference x rotational speed
    = $2\pi r$ x n
    = $2\pi rn$

So substituting $2\pi rn$ for v in the formula $F = mv^2/r$ gives

$$F_c = \frac{m(2\pi rn)^2}{r}$$

or

$$F_c = \frac{mr^2(2\pi n)^2}{r}$$

and

$$F_c = mr(2\pi n)^2$$

This formula shows that centrifugal force is proportional to both the mass and the radius, but the force varies with the square of the rotational speed, e.g. the force at 20 rev/s is four times that at 10 rev/s. Due to the rapid increase of force with speed, the balance of rotating parts is essential.

Balancing

Many motor vehicle components rotate at high speed, so severe vibration occurs if any unbalance exists. Some components, such as road wheels, are balanced by a mechanic, but many others are only balanced during manufacture. In the latter case, the mechanic should ensure that any part removed from a component is refitted in its original position.

Correction of out-of-balance components is achieved by either removing material from the heavy part or by adding material on the opposite side to the heavy spot, so as to give a counterbalance action (Figure 2.23.3).

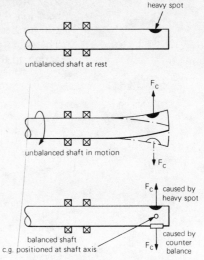

Figure 2.23.3 Balancing a shaft

When the force given by the counterbalance does not act in the same plane, another form of vibration occurs. In Figure 2.23.4 the wheel has good 'static balance' but when it is rotated, the heavier parts exert a centrifugal force which produces an unbalanced couple.

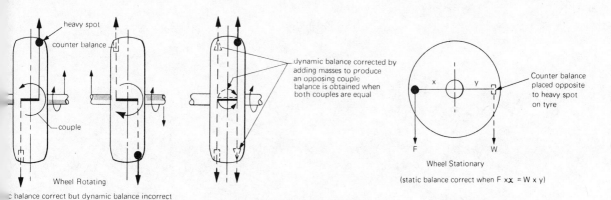

Figure 2.23.4 A balanced road wheel

To correct this, material must be added to produce an equal and opposite couple. Incorrect dynamic balance is particularly important in respect to road wheels, since the steering layout will allow the wheel to vibrate about the swivel axis or king pin at a frequency equal to the speed of wheel rotation. As stated previously, the force causing this disturbance greatly increases with speed, so at high vehicle speeds a severe 'flapping' of the road wheel (wheel shimmy) and oscillation of the steering wheel is experienced.

Dynamic balance also applies to crankshafts. The crankshaft shown in Figure 2.23.5(a) is not used because the couples are un-balanced — both tend to cause the engine to move in the direction shown. Rearranging the crank pins to the form indicated in Figure 2.23.5(b) improves the engine balance, and this is the arrangement used in a 4-cylinder in-line engine. This type of shaft can be improv[ed] still further by incorporating counterbalance masses (Figure 2.23.5[c]). These masses produce opposing couples which relieve both the stre[ss] acting on the shaft and the load taken by the centre main bearing.

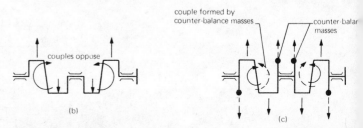

Figure 2.23.5 Crankshaft balance

Stability

Occasions sometimes arise where a vehicle is jacked up and the risk [of] overturning is present. An object overturns or 'falls over' when the vertical line through the object's centre of gravity falls outside its base.

Figure 2.23.6 shows a loaded vehicle with a high centre of gravi[ty]. In the position shown the vehicle is on the verge of overturning so action such as jacking-up the right-hand side will have an obvious result.

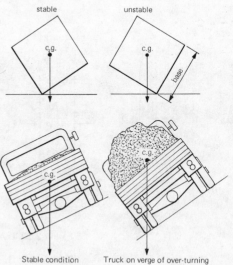

Figure 2.23.6 Stability

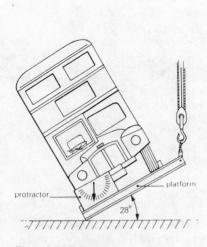

Figure 2.23.7 Tilting test to check stability of a bus

A p.s.v., such as the double-decker bus shown as Figure 2.23.7 must pass a stability or tilt test to ensure that the vehicle with th[e] upper deck laden does not overturn when it is tilted to a specified angle.

.24 Machines

Previous work on machines introduced the following:
The input force applied is called the *effort*.
The output force from the machine is the *load*.

$$\text{Movement ratio} = \frac{\text{distance moved by effort}}{\text{distance moved by load}}$$

$$\text{Force ratio} = \frac{\text{load}}{\text{effort}}$$

$$\text{Efficiency} = \frac{\text{work output}}{\text{work input}} \text{ OR } \frac{\text{force ratio}}{\text{movement ratio}}$$

As applied to gearboxes these expressions can also be stated as

$$\text{Movement ratio} = \text{gear ratio} = \text{speed ratio}$$

$$\text{Force ratio} = \text{torque ratio} = \frac{\text{output torque}}{\text{input torque}}$$

$$\text{Efficiency} = \frac{\text{torque ratio}}{\text{gear ratio}}$$

Example 1

Figure 2.24.1 shows the layout of a gearbox. Calculate
(a) the gear ratio
(b) the output torque if the efficiency is 75 per cent

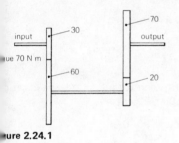

input

output

—70

—30

ue 70 N m

—60

—20

ure 2.24.1

(a)
$$\text{Gear ratio} = \frac{\text{driven}}{\text{driver}} \times \frac{\text{driven}}{\text{driver}}$$
$$= \frac{60}{30} \times \frac{70}{20}$$
$$= 7 : 1$$

(b)
$$\text{Efficiency} = \frac{\text{torque ratio}}{\text{gear ratio}}$$

∴
$$\text{torque ratio} = \text{efficiency} \times \text{gear ratio}$$
$$= 0 \cdot 75 \times 7$$
$$= 5 \cdot 25$$

since
$$\text{torque ratio} = \frac{\text{output torque}}{\text{input torque}}$$

then
$$\text{output torque} = \text{torque ratio} \times \text{input torque}$$
$$= 5 \cdot 25 \times 70$$
$$= 367 \cdot 5 \text{ N m}$$

Part (b) could also be solved by

Output torque (when efficiency is 100%) = input torque × gear ratio
$$= 70 \times 7$$
$$= 490 \text{ N m}$$

Output torque (when efficiency is 75%) = 490 x 0·75

= 367·5 N m

Example 2

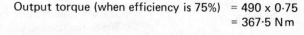

A two-stage chain system is used to drive the two camshafts shown in Figure 2.24.2. Calculate

(a) the number of teeth on the smaller intermediate sprocket

(b) the torque ratio if the drive is 90 per cent efficient

(a) The camshafts must be driven at half crankshaft speed.

∴ ratio = 2 : 1

$$\text{gear ratio} = \frac{\text{driven}}{\text{driver}} \times \frac{\text{driven}}{\text{driver}}$$

Let x represent the number of teeth on the smaller intermediate sprocket

$$\text{then } 2 = \frac{28}{21} \times \frac{30}{x}$$

$$2 = \frac{28 \times 30}{21x}$$

$$42x = 28 \times 30$$

$$x = \frac{28 \times 30}{42} = 20 \text{ teeth}$$

(b) If efficiency is 100 per cent, then torque ratio = speed ratio

However, since efficiency is only 90 per cent

$$\text{torque ratio} = 2 \times 0·9$$

$$= 1·8$$

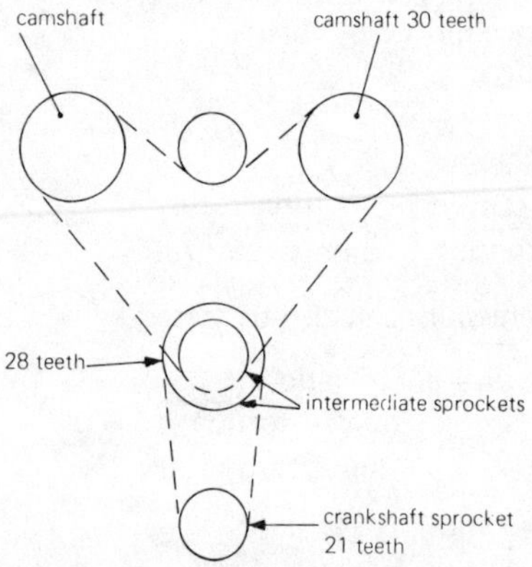

camshaft

camshaft 30 teeth

28 teeth

intermediate sprockets

crankshaft sprocket
21 teeth

Figure 2.24.2

Example 3

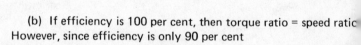

At an engine speed of 2025 rev/min, a torque of 96 N m is supplied to the gearbox. If the gearbox and final drive ratios are 2·5 and 4· respectively, calculate

(a) the overall gear ratio

(b) the torque supplied by the differential to each road wheel (frictional losses may be neglected)

(c) the speed of the road wheel

(a) Overall gear ratio is the *product* of gearbox ratio and final drive ratio.

$$\text{Overall gear ratio} = 2·5 \times 4·5$$

$$= 11·25$$

(b) Since efficiency is 100 per cent

then torque ratio = gear ratio

so total torque output from differential = input torque

x torque ratio

= 96 x 11·25

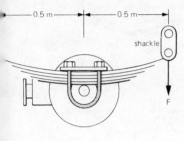

The differential divides the torques equally between each driving wheel irrespective of speed, therefore

$$\text{torque applied to each wheel} = \frac{1080}{2} = 540 \, \text{Nm}$$

(c) Speed of road wheel $= \dfrac{\text{engine speed}}{\text{combined gear ratio}}$

$$= \frac{2025}{11 \cdot 25} \approx 180 \, \text{rev/min}$$

—0.5 m— —0.5 m—

shackle

F

Example 4

Gearbox and final drive ratios of a vehicle are 4:1 and 5:1, respectively. If the engine torque is 150 Nm, and friction losses are neglected, determine

    (a) the torque reaction of the rear axle

    (b) the force applied to each spring shackle of the torque reaction

if the spring dimensions are as shown in Figure 2.24.3

Figure 2.24.3

    (a) Torque reaction of axle is equal and opposite to torque applied to axle shafts.

∴   Torque reaction = engine torque x combined gear ratio
$$= 150 \times 20$$
$$= 3000 \, \text{N m}$$

    (b) Torque reaction is taken by the spring mounting points.

Torque = force x radius

so total force acting on the four mounting points $= \dfrac{\text{torque}}{\text{radius}}$

$$= \frac{3000}{0 \cdot 5} = 6000 \, \text{N}$$

∴   force applied to one spring shackle $= \dfrac{6000}{4} = 1500 \, \text{N}$

Multi-drive rear axles

The previous example serves as an introduction to the suspension arrangements for rigid six-wheeled vehicles. Mounting semi-elliptic springs in the manner shown in Figure 2.24.4 causes the axles to be overloaded, so various interconnections are made between the spring in order that each axle will share the load.

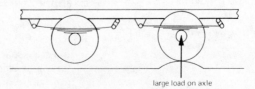

large load on axle

Figure 2.24.4 Suspension system which does not distribute the load

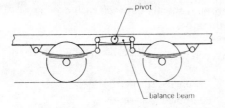

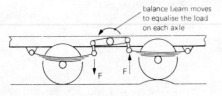

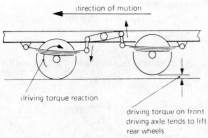

Figure 2.24.5 Heavy commercial vehicle suspension system

Figure 2.24.5 shows one simple method of dividing the axle load but this system is far from ideal, because the forces produced by torque reaction tend to lift one axle from the road.

Modern vehicles use layouts similar to Figure 2.24.6 and these are designed to

(a) divide the suspension load equally

(b) absorb the torque reaction of each driving axle

(c) transmit the driving thrust from the wheel to the frame

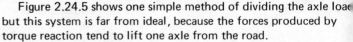

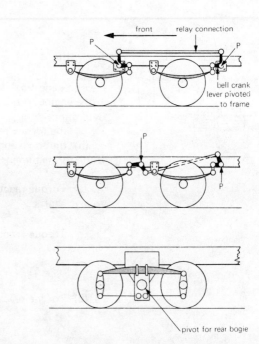

Figure 2.24.6 C.V. suspension systems

Belt drives

The belt is still the most common method used for driving external engine auxiliaries such as the alternator, water pump, fan, pump for power steering, etc. During the past few years the introduction of the toothed belt has extended the application to include drives to camshafts and other components which demand non-slip features.

The vee belt relies on its wedging action (Figure 2.24.7) to provide the grip. This allows the torque applied by the driving pulley to produce a tensile force in the belt which in turn applies a torque to the driven pulley. Belt and chain drives are examples where rotary motion is converted into linear motion.

Example 5

A crankshaft pulley and alternator pulley have effective radii of 60 mm and 48 mm respectively (Figure 2.24.8). If the torque applied by the crankshaft is 1·5 N m at a speed of 49 rev/s calculate

(a) the tensile force in the belt due to the torque

(b) the linear speed of the belt

(c) the torque and speed of the alternator pulley (frictional loss may be neglected)

Figure 2.24.7 'V' belt drive

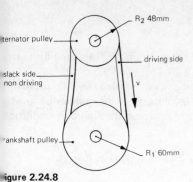

Figure 2.24.8

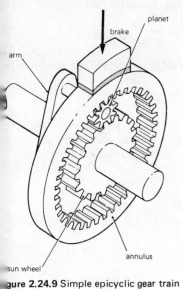

Figure 2.24.9 Simple epicyclic gear train

(a)
$$\text{torque} = \text{force} \times \text{radius}$$

$$\text{tensile force in belt} = \frac{\text{torque at crankshaft pulley (Nm)}}{\text{effective radius of crankshaft pulley (m)}}$$

$$= \frac{1 \cdot 5}{0 \cdot 06} = 25 \text{ N}$$

(b) linear speed of belt = linear speed of pulley at radius of 60 mm
= circumference × speed of rotation
$$= 2\pi r_1 n$$
$$= 2 \times \frac{22}{7} \times 0 \cdot 06 \times 49$$
$$= 18 \cdot 48 \text{ m/s}$$

(c) torque on alternator pulley = force × radius
$$= 25 \times 0 \cdot 048$$
$$= 1 \cdot 2 \text{ Nm}$$

An easy method for finding the speed of the driven pulley is to apply the same rule as that used for gearing

$$\text{movement ratio} = \text{speed ratio} = \frac{\text{driven}}{\text{driver}}$$

In this case the radii is used instead of numbers of teeth.

so
$$\text{speed ratio} = \frac{\text{radius of driven pulley}}{\text{radius of driving pulley}}$$

$$= \frac{48}{60} = 0 \cdot 8 : 1$$

∴
$$\text{speed of alternator pulley} = \frac{49}{0 \cdot 8} = 61 \cdot 25 \text{ rev/s}$$

Epicyclic gearing

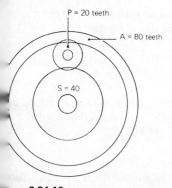

Figure 2.24.10

A small gear rolling on the circumference of another gear is the meaning of the term epicyclic. This type of gearing is commonly used in automatic gear boxes; the gear selection is obtained by applying a band brake or by locking a multi-disc clutch. Connecting the input shaft to different gear elements by clutches enables a number of different ratios to be obtained from one epicyclic gear train. When the advantages of this compact layout is linked with the ease and speed that the gear change can be effected, then the reasons for the use of this type of gearing will be apparent.

Figure 2.24.9 shows the main components of a simple epicyclic gear train; the layout shown is that used when a large speed reduction is required.

Gear ratio

The gear ratio of a train arranged as a Figure 2.24.10 can be calculated from the formula

$$\text{Ratio} = \frac{A + S}{S}$$

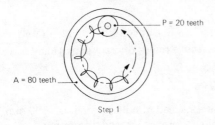

Step 1

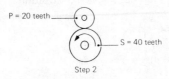

Step 2

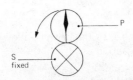

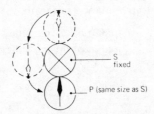

P will rotate twice when it is rotated about S

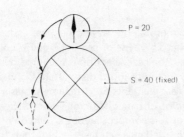

P will rotate three times when it is rotated about S

Step 3

Figure 2.24.11

Hydraulic machines

where A = teeth on annulus

S = teeth on sun

This formula can be proved by applying the following three steps to the layout shown in Figure 2.24.11

Step 1: The ratio is the number of turns made by the sun in order to rotate the arm through one revolution, so the number of turns made by the planet when the arm is moved through one revolution is as follows

$$\text{number of turns} = \frac{80}{20} = 4 \text{ times OR } \frac{A}{P} \text{ times}$$

Step 2: To rotate the planet one revolution, the sun must move

$$\frac{20}{40} = \frac{1}{2} \text{ revolutions OR } \frac{P}{S} \text{ revolution}$$

so to rotate the planet 4 times (or A/P times), the sun must rotate

$$4 \times \frac{1}{2} = 2 \text{ times OR } \frac{A}{R} \times \frac{R}{S} = \frac{A}{S} \text{ times}$$

Step 3: The point of contact between sun and planet passes through 360° or one revolution as the planet revolves around the sun, so the 'extra one' must be added to the result obtained in Step 2.

$$\therefore \qquad \text{Ratio} = \frac{A}{S} + 1$$

$$= \frac{A}{S} + \frac{S}{S}$$

$$= \frac{A + S}{S} = 3 : 1$$

(This 'extra one' can be demonstrated with two coins.)

Only in cases where the planet revolves around the sun does the 'extra one' have to be added.

Rearranging the layout by connecting the input shaft to the annulus and applying a brake to the sun, gives the ratio: $(A + S)/A$. Using the same gear sizes as in Figure 2.24.10 will give a ratio of

$$\frac{80 + 40}{80} = 1.5 : 1$$

Figure 2.24.12 shows the variations which may be obtained by connecting up an epicyclic train in different ways. For further examples of epicyclic gearing see *F. of M.V.T.*

A hydraulic machine is a device or system in which a liquid is used to transmit energy. The motor vehicle uses many hydraulic machines, these include systems for operating, or assisting in the operation of the brakes, steering, automatic gearboxes, suspension and many power take-off auxiliaries.

Figure 2.24.13 shows the layout and principle of a simple hydraulic machine.

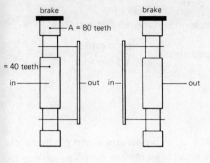

3:1 Reduction   1:3 Overdrive

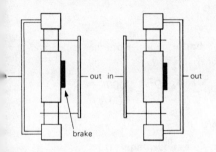

1·5:1 Reduction   1:1·5 Overdrive

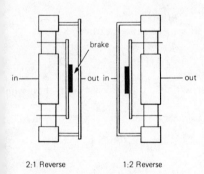

2:1 Reverse   1:2 Reverse

2.24.12

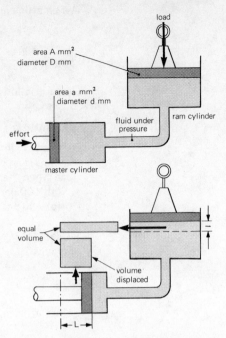

Figure 2.24.13 Simple hydraulic machines

The fluid displaced (pumped) from the master cylinder is used to move the ram, so if the latter has a larger diameter, its movement will be less. Therefore.

$$\text{movement ratio} = \frac{\text{movement of effort}}{\text{movement of load}} = \frac{L}{\ell}$$

Since
  fluid displaced by master cylinder = fluid received by ram cylinder
then
  area of master cylinder plunger x stroke =
              area of ram cylinder piston x ram stroke

or $$a \times L = A \times \ell$$

and $$\frac{L}{\ell} = \frac{A}{a}$$

so $$\text{movement ratio} = \frac{L}{\ell}$$

$$= \frac{A}{a}$$

$$= \frac{\pi D^2/4}{\pi d^2/4} = \frac{D^2}{d^2}$$

$$\text{Force ratio} = \frac{\text{load}}{\text{effort}}$$

When efficiency is 100 per cent

$$\text{force ratio} = \text{movement ratio}$$

so
$$\text{force ratio} = \frac{\text{area of ram}}{\text{area of plunger}}$$

when friction is neglected.

Example 6

A master cylinder of diameter 30 mm operates a ram of diameter 60 mm. If an effort of 80 N acts at the master cylinder, what is the load when the efficiency is
(a) 100 per cent
(b) 90 per cent

(a) Let d = diameter of master cylinder plunger (mm) and D = diameter of ram piston (mm).

$$\text{Movement ratio} = \frac{D^2}{d^2} = \frac{3600}{900} = 4$$

Since efficiency is 100 per cent

$$\text{force ratio} = \text{movement ratio}$$
$$\therefore \quad \text{force ratio} = 4$$
$$\text{force ratio} = \frac{\text{load}}{\text{effort}}$$
$$\therefore \quad \text{load} = \text{force ratio} \times \text{effort}$$
$$= 4 \times 80$$
$$= 320 \text{ N}$$

(b) If load is 320 N when efficiency is 100 per cent, then when efficiency is 90 per cent

$$\text{load} = 320 \times \frac{90}{100} = 288 \text{ N}$$

Hydraulic operation is particularly suited for braking systems. Low friction losses in the system combined with the ease in which a hydraulic supply can be applied to the brake cylinders are major advantages. Furthermore, the fact that pressure cannot be applied to any brake until all clearances have been taken-up simplifies the task of brake adjusting and ensures that each brake starts to apply at the same time. Also, when the brakes do act, the pressure received by each one is the same, i.e. the self-compensating feature ensures that each brake receives its share of the applied effort. This feature can also cause problems; any leakage in the system will prevent any of the brakes from functioning, so to avoid this, a tandem cylinder or dual layout is used.

A hydraulic machine will not function efficiently if air is present in the system. Air, unlike liquids, is compressible so the effect, as applied to a braking system, is similar to that produced when a spring is placed between the driver's foot and the brake pedal.

The action of a brake system may be considered by referring to Figure 2.24.14.

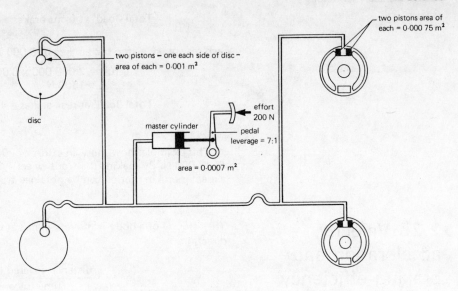

two pistons area of each = 0·000 75 m²

two pistons – one each side of disc – area of each = 0·001 m²

disc

effort 200 N

master cylinder

pedal leverage = 7:1

area = 0·0007 m²

Figure 2.24.14 Layout of hydraulic brake system

When friction is neglected

Pedal effort = 200 N

Force applied to master cylinder

= 200 x leverage

= 200 x 7

= 1400 N

Area of master cylinder plunger

= 700 mm²

= 0·000 7 m² (since there are 1000 x 1000 mm² = 1 m²)

Pressure in system

$$= \frac{\text{force (N)}}{\text{area (m}^2)}$$

$$= \frac{1400}{0·0007} = \frac{14\,000\,000}{7} \text{ N/m}^2$$

= 2 000 000 N/m² OR 20 bar
(since 1 bar = 100 000 N/m² OR 10^5 N/m²)

This pressure is transmitted throughout the system.
The thrust on each wheel cylinder piston is given by

thrust = pressure x area

Each front brake piston has an area of 0·001 m²

∴   thrust = 2 000 000 x 0·001
= 2000 N

Total 'load' at front brakes = 4 x 2000
= 8000 N

Each rear brake piston has an area of 0·000 75 m^2

∴ thrust = 2 000 000 x 0·000 75
= 1500 N

. Total 'load' at rear brakes = 4 x 750
= 6000 N

This example shows how an effort of 200 N can produce a 'loa of 14 000 N. By making the front wheel cylinders larger than the a greater braking action can be obtained from the front wheels.

2.25 Velocity, acceleration and braking efficiency

The velocity of a body is the rate in m/s at which it moves in a giv direction.

$$velocity = \frac{distance\ covered\ (metres)}{time\ (seconds)}$$

Velocity

or

$$v = \frac{s}{t}$$

The expression 'distance covered/time' is also used for determi the *speed* of a body; the difference between the terms speed and velocity is that speed does not consider the direction of motion.

Acceleration

Acceleration is the rate of increase in velocity and is expressed in metres per second squared (m/s^2). The velocity-time graph shown in Figure 2.25.1 represents a vehicle accelerating from rest at a constant rate of 2 m/s^2. It shows that after each second of time t velocity increases by the amount a — in this case 2 m/s.
Over a period of 3 seconds the

increase in velocity = 6 m/s
or acceleration = 6 m/s per 3 s
= 6 m/s/3 s
= 2 m/s/s OR 2 m/s^2

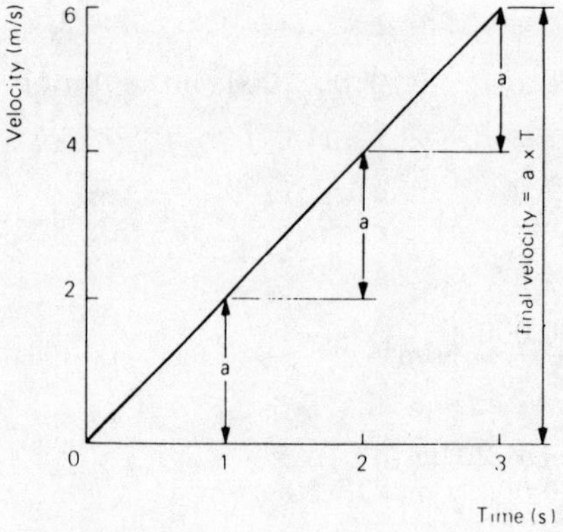

From the graph it will be seen that

final velocity = acceleration x time
$v = a \cdot t$

Figure 2.25.1

where v is in m/s
a is in m/s^2
t is in s

The force of gravity causes an object to accelerate at a constan rate which has a value that varies slightly in different parts of the world. It has been agreed internationally that the average value should be 9·806 65 m/s^2.

For general work this is approximated to either 9·81 m/s^2 or 10 m/s^2 — the value used depends on the accuracy required.

When a motor vehicle is slowed down, the velocity decreases. This negative acceleration is called deceleration: the instrument used to measure this rate of retardation is called a *decelerometer.*

king efficiency

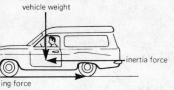

re 2.25.2 Brake efficiency is 100 per cent when

etarding force = Total vehicle weight

(inertia force is always equal to the retarding force)

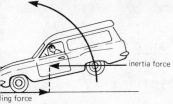

re 2.25.3 Couple produced by the two forces gives an overturning action

re 2.25.4 Force of impact can cause a deceleration in excess of 30 *g*

etic energy

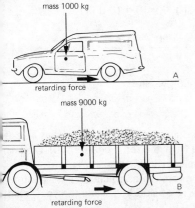

re 2.25.5 Stopping distance of vehicle B is similar to that of vehicle A because retarding force is pro- portional to mass

Gravitational acceleration or *g* can be used as the standard for brake efficiency. If a vehicle on a level road was decelerated by the brakes at the rate of 9·81 m/s^2, the braking efficiency is said to be 100 per cent.

100 per cent brake efficiency = declaration at the rate of g

To produce an efficiency of 100 per cent, the brakes must give, at the road surface, a retarding force equal to the total weight of the vehicle; this would require a coefficient of friction between the tyre and road of 1·0 (Figure 2.25.2). Since this friction value is far above the average, an efficiency higher than 90 per cent is seldom achieved.

100 per cent brake efficiency

$=$ **brakes apply a retarding force equal to the total weight of the vehicle including its load**

Statutory regulations specify the minimum efficiency for each type of vehicle, e.g. a four-wheeled car: foot brake 50 per cent, hand brake 25 per cent.

Figure 2.25.3 shows that a deceleration of the vehicle causes an inertia force which is equal to the retarding force. The combined effect of these forces causes a weight transference from the rear to the front and for this reason the front brakes have a larger braking capacity.

The force of impact during a collision (Figure 2.25.4) can cause a deceleration in excess of 30 *g*. Subjecting a driver to high decelerations of this order produces a force on his body greater than thirty times his weight. Special chassis constructions, which are designed to give a concertina effect on impact, are intended to decrease the deceler- ation felt by the vehicle's occupants and so reduce the risk of injury.

A braking system converts kinetic energy (the energy of motion) to heat energy. The rate at which this energy is converted governs the distance required to bring the vehicle to rest
Kinetic energy can be calculated from

$$KE = \tfrac{1}{2}\,mv^2$$

where KE = energy (J)
m = mass (kg)
v = velocity (m/s)

When a vehicle is moving at a given velocity then

kinetic energy = work stored

$=$ retarding force x stopping distance

$\therefore$ $\quad$ stopping distance $= \dfrac{\text{kinetic energy}}{\text{retarding force}}$ OR $\dfrac{\tfrac{1}{2}\,mv^2}{F}$

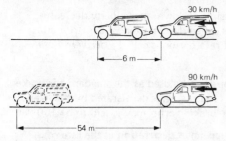

Figure 2.25.6 Stopping distance varies as the square of the velocity

This expression enables the following conclusions to be formed:

1. the stopping distance is governed by the ratio of the retardir force to the vehicle mass (Figure 2.25.5).

2. the stopping distance varies as the square of the velocity, e.g the distance required to stop a vehicle travelling at 90 km/h is nine times as great as that required for a speed of 30 km/h (Figure 2.25

Example 1

A retarding force of 5 kN acts on a vehicle of mass 1 tonne (1000 which is moving at 72 km/h. Calculate the

(a) stopping distance

(b) brake efficiency. (Take g as 10 m/s^2).

(a)
$$72 \text{ km/h} = \frac{72\,000}{60 \times 60} \text{ m/s}$$

$$= 20 \text{ m/s}$$

$$KE = \tfrac{1}{2} mv^2$$

$$= \frac{1000 \times 20^2}{2}$$

$$= \frac{1000 \times 400}{2}$$

$$= 200\,000 \text{ J}$$

$$\text{stopping distance} = \frac{KE}{\text{retarding force}}$$

$$= \frac{200\,000}{5000}$$

$$= 40 \text{ m}$$

(b)
$$\text{Brake efficiency} = \frac{\text{retarding force}}{\text{total weight of vehicle}} \times \frac{100}{1}$$

Mass is 1000 kg, so weight = 1000 x g

$$= 1000 \times 10 = 10\,000 \text{ N}$$

so
$$\text{efficiency} = \frac{5000}{10\,000} \times \frac{100}{1}$$

$$= 50 \text{ per cent}$$

2.26 Exercises and multiple choice questions

Exercises

1. The force required to drag a vehicle with all wheels skiddin along a level surface is 1200 N. If the force between the tyres anc road is 2000 N, calculate the coefficient of friction.

2. Calculate the retarding force of a vehicle of mass 1 tonne i all wheels are held on the verge of skidding. Take the coefficient of friction as 0·55 and g as 10 m/s^2.

3. A brake pad is pressed against a disc by a force of 80 N. If the pad acts at a radius of 180 mm and the coefficient of friction is 0·3, calculate the braking torque given by the pad.

4. A brake shoe pressed against a drum by a force of 600 N produces a drag of 210 N on the drum. What is the coefficient of friction between the lining and drum.

5. Water entering a brake drum lowers the coefficient of friction by 80 per cent. What braking torque will be obtained from water-soaked linings if the torque given by a dry brake is 700 Nm?

6. A single plate clutch has a mean radius of 200 mm, total spring thrust of 960 N and coefficient of friction of 0·3. Calculate the torque transmitted by this clutch.

7. A clutch transmits a torque of 280 Nm at a speed of 2400 rev/min. Calculate the power transmitted.

8. What spring force is necessary to transmit a torque of 54 Nm by a single-plate clutch having a mean radius of 100 mm and coefficient of friction of 0·3?

9. A force of 35 N is applied at right angles to the top end of a straight level of total length 300 mm. What force is given at the bottom end of the lever if the pivot is 50 mm from the lower end?

10. What is the magnitude of a couple formed by two forces of 40 N which act in parallel at a distance of 80 mm apart?

11. A mass of 5 kg is attached to the end of a horizontal lever of total length 300 mm. What force must be applied vertically to the other end of the lever if the pivot is situated 200 mm from the 5 kg mass. (Take g as 10 m/s^2.)

12. Figure 2.26.1 shows the layout of a system of levers with the pivots marked as P_1 and P_2. What force is produced by the rod marked A? (Friction may be neglected).

13. The rear wheels of a vehicle of mass 10 tonne support a load of 40 kN. If the wheelbase is 5 m, calculate the position of the centre of gravity. (Take g as 10 m/s^2.)

14. A vehicle of mass 1200 kg has a wheelbase of 2·5 m. Calculate the position of the centre of gravity if the load on the front wheels is 9·6 kN. (Take g as 10 m/s^2.)

15. When a connecting rod is positioned horizontally and supported by spring balances positioned at the centres of the small-end and big-end bearings, the readings on the balances are 6 N and 18 N, respectively. Calculate the position of the centre of gravity if the length of the rod, measured between centres is 240 mm.

16. A gearbox gear train is shown in Figure 2.26.2. Calculate the speed and torque at the output shaft if the efficiency is 80 per cent.

17. A two-stage chain camshaft drive has sprockets of the following sizes

crankshaft and small intermediate — 25 and 40 teeth
large intermediate   — 48 teeth

Calculate the number of teeth on the camshaft sprocket assuming a chain connects the large intermediate sprocket to the camshaft.

18. At an engine speed of 3675 rev/min a torque of 150 N m is supplied to the gearbox. If the gearbox and final drive ratios are 2·5 and 4·2 respectively and the vehicle is travelling in a straight line, calculate the

(a) speed of each road wheel
(b) torque applied to each road wheel if the efficiency is 100 per cent.

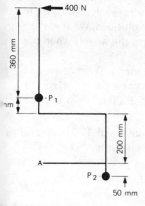

re 2.26.1

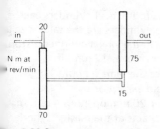

re 2.26.2

19. A transmission system has an overall ratio of 4·4 : 1 and wh the engine torque is 70 Nm the torque applied to each road wheel is 115·5 Nm. Calculate the efficiency.

20. A crankshaft pulley and alternator pulley have effective rad of 60 mm and 50 mm respectively. If the torque applied by the cra shaft is 2·4 Nm at a speed of 56 rev/s calculate the

(a) tensile force in the belt due to the torque

(b) linear speed of the belt

(c) torque and speed of the alternator pulley

(Neglect friction losses.)

21. With second gear engaged 10 revolutions of the crankshaft turns the crown wheel of the final drive one revolution. If the fina drive ratio is 4 : 1, calculate the speed of the propeller shaft when second gear is engaged and the engine speed is 3500 rev/min.

22. During a speedometer check one driving wheel was lifted from the ground, and when this wheel was revolved 6 times the speedometer cable turned through 9 revolutions. What is the gear ratio between the speedometer cable and the wheels when the vehicle is travelling on a straight path?

23. A crankshaft vee pulley of diameter 120 mm drives a dynamo pulley of diameter 140 mm. What is the speed of the dynamo armature at an engine speed of 3500 rev/min assuming 10 per cent slip?

24. An epicyclic gear train has an annulus of 100 teeth and su gear of 40 teeth. What is the gear ratio when the

(a) sun is driving and the annulus is held

(b) annulus is driving and the sun is held

(c) sun is driving and the arm (planet carrier) is held

25. A master cylinder of area 590 mm^2 acts on a ram cylinder area 14 750 mm^2. What force will be given by the ram if the effor applied to the master cylinder is 400 N and the efficiency is 90 pe cent?

26. A master cylinder of area 600 mm^2 operates two slave cylinders A and B; cylinder A has an area of 1800 mm^2 and cylinder B an area of 2100 mm^2. What force is given by each cylinder when the effort at the master cylinder is 250 N and the efficiency is 100 per cent?

27. A hydraulic brake system has four wheel cylinders each containing two pistons and all wheel cylinders are the same size a the master cylinder. What distance will the wheel cylinder piston move when the master cylinder piston moves 12 mm, if

(a) each wheel cylinder piston moves the same amount

(b) all wheel cylinder pistons are held except one

28. A master cylinder of diameter 20 mm operates a ram of diameter 50 mm. If an effort of 60 N acts at the master cylinder, what is the load when the efficiency is 90 per cent.

29. A vehicle accelerates from rest at the rate of 1·5 m/s^2. Wh is its velocity after 8 seconds?

30. Travelling at a constant speed a vehicle covers a distance o 324 m in 12 seconds. What is its velocity?

31. A vehicle is brought to rest from a velocity of 52 m/s in 2 seconds. What is the average deceleration?

32. What is the brake efficiency of a vehicle which is brought rest from a velocity of 105 m/s in 15 seconds. (Take *g* as 10 m/s^2

33. How long will it take to stop a vehicle from a speed of 86·4 km/h if the brake efficiency is 60 per cent? (Take g as 10 m/s^2.)

34. To stop a vehicle from a speed of 86·4 km/h takes 4 seconds. If the average velocity during this period as 12 m/s, find the stopping distance.

35. A retarding force of 9·75 kN is applied by the brakes in stopping a vehicle of mass 1500 kg. Calculate the brake efficiency. (Take g as 10 m/s^2.)

36. A retarding force of 9 kN acts on a vehicle of mass 1200 kg, which is moving at 108 km/h. Taking g as 10 m/s^2, calculate the
(a) kinetic energy
(b) stopping distance
(c) brake efficiency

Multiple-choice questions

1. When one surface rubs against another the resistance to sliding is called
   (a) retardation
   (b) absorption
   (c) friction
   (d) fade

   (a) π
   (b) μ
   (c) η
   (d) θ

2. The expression 'static friction' means the
   (a) resistance to motion
   (b) resistance to separate two surfaces
   (c) force which resists motion when one surface is sliding over another at a constant speed
   (d) force which opposes the initial movement of one surface relative to another

3. If F is the force needed to overcome friction and W is the force pressing the surfaces together, then coefficient of friction is
   (a) $W - F$
   (b) $W + F$
   (c) $F \times W$
   (d) F/W

4. Coefficient of friction is normally represented by the symbol

5. A typical value for the coefficient of friction of a rubber tyre on a dry road surface is
   (a) 0·2
   (b) 0·6
   (c) 1·0
   (d) 1·4

6. A block sliding over a horizontal surface has a coefficient of friction of 0·4. If the friction area is doubled, the coefficient of friction will be
   (a) 0·2
   (b) 0·4
   (c) 0·8
   (d) 1·6

7. Friction between two rubbing surfaces is *increased* when the
   (a) contact area is decreased
   (b) speed of rubbing is decreased
   (c) surfaces are lubricated
   (d) surfaces are roughened

101

8. Friction between two rubbing surfaces is *decreased* when the
   (a) contact area is increased
   (b) speed of rubbing is increased
   (c) surfaces are lubricated
   (d) surfaces are roughened

9. How does the energy required to rotate a shaft supported on plain bearings compare with that required to rotate a similar shaft which uses roller bearings? The effort needed to rotate the shaft in the plain bearings is
   (a) smaller, because lubrication is better
   (b) smaller, because the contact area is larger
   (c) greater, because the contact area is larger
   (d) greater, because rolling friction is less than sliding friction

10. What is the effect of increasing the friction area of a disc brake pad?
    (a) The friction will be increased
    (b) The coefficient of friction will be increased
    (c) The rate of wear will decrease
    (d) The pad will operate at a higher temperature

11. Heating a friction brake lining to a temperature greater than 300 °C causes the
    (a) lining to grab
    (b) material to 'fade'
    (c) coefficient of friction to increase
    (d) asbestos to weld to the metal surface

12. A friction force of 24 N is required to slide one surface over another. If the force pressing the surfaces together is 75 N the coefficient of friction
    (a) 0·18
    (b) 0·3125
    (c) 0·32
    (d) 0·385

13. Two surfaces having a coefficient of friction of 0·2 are pressed together with force of 50 N. The friction force is
    (a) 12 N
    (b) 24 N
    (c) 48 N
    (d) 208 N

14. A frictional force of 32 N is needed to slide a block mass 8 kg along a horizontal surface. When g is taken a 10 m/s^2, the coefficient o friction is
    (a) 0·25
    (b) 0·256
    (c) 0·32
    (d) 0·4

15. What is the maximum retarding force of a vehicl of mass 900 kg when the coefficient of friction is 0·6?
    Taking g as 10 m/s^2 the retarding force is
    (a) 0·54 kN
    (b) 1·5 kN
    (c) 5·4 kN
    (d) 15 kN

16. A lubricant reduces fricti by
    (a) absorbing the fricti energy
    (b) applying pressure ir the direction of motion

(c) polishing the two surfaces

(d) holding the surface apart

17. The property of a lubricant to 'cling to a metal surface' is called
    (a) oiliness
    (b) friction
    (c) viscosity
    (d) stiction

18. The expression 'boundary lubrication' is used to describe a condition where an oil film is maintained by the
    (a) viscosity property of the lubricant
    (b) oiliness property of the lubricant
    (c) pressure supplied by an oil pump
    (d) force which pushes the two metal surfaces together

19. A Redwood viscometer is an instrument used to measure a lubricant's
    (a) density
    (b) resistance to flow
    (c) resistance to wear
    (d) ability to cling to a metal surface

20. The viscosity of an oil is indicated by the
    (a) SAE number
    (b) viscosity index
    (c) temperature change
    (d) temperature index

21. An oil having a high viscosity index is one in which the viscosity
    (a) increases as the temperature increases
    (b) remain constant over a large temperature range

(c) varies relatively little with change in temperature

(d) alters a large amount if the temperature is changed

22. To obtain a standard, Pennsylvanian oil and Gulf Coast oil are given values of 100 and 0 respectively. Information on the behaviour of these oils is used to obtain a lubricant's
    (a) viscosity index
    (b) SAE rating
    (c) Redwood number
    (d) resistance to flow

23. An oil classed as 'multi-grade' or 'cross-grade' is one which
    (a) gets thicker with increase in temperature
    (b) maintains a constant viscosity when heated
    (c) has a low viscosity index
    (d) has a high viscosity index

24. Compared to an oil rated as SAE 40, one advantage of using the multigrade oil SAE 20W 50 in an engine is that the multi-grade oil
    (a) is thinner at high temperatures
    (b) is thicker at low temperatures
    (c) does not alter its viscosity
    (d) can be used through-out the year

25. Grease is a substance which
consists of
    (a) a lime and soap
solution
    (b) lime-soap, bentone
and lithium
    (c) a mixture of low
viscosity lubricants
    (d) heavy oil mixed with
a retaining base

26. What effect will result from
a fault which causes the output shaft of a friction clutch
to rotate slower than the
input shaft?
    (a) Heat will be generated
    (b) Extra torque will be
transmitted
    (c) Engine speed will
decrease
    (d) Mechanical efficiency
will increase

27. The torque transmitted by
a plate type clutch is given
by $T = S p \mu r$. During the
life of the clutch which
two factors in the expression
$S p \mu r$ decrease in value
and cause 'slip'
    (a) S and p
    (b) p and μ
    (c) μ and r
    (d) r and S

28. A single plate clutch of mean
radius 120 mm has a total
spring thrust of 500 N and
coefficient of friction of 0·3.
The torque transmitted is
    (a) 18 N m
    (b) 36 N m
    (c) 18 kN m
    (d) 36 kN m

29 Partial seizure of a clutch
pressure plate in the cover
prevents 30 per cent of the
spring thrust from acting on
the driven plate. The effect
of this fault is that the
clutch will

(a) drag
(b) spin
(c) transmit only 70 per
cent of its original
torque capacity
(d) transmit only 30 per
cent of its original
torque capacity

30. Excessive use of a clutch
has heated the linings and
caused 'fade'. The effect
of this is that
    (a) drag will occur
    (b) spin will occur
    (c) the maximum torque
capacity of the clutch
is reduced
    (d) the maximum torque
capacity of the clutch
is increased

31. A clutch driven plate is
replaced by a plate which
has the same outside dia
meter but the inner dia-
meter of the linings is
much smaller than that
used on the original plate
A possible effect which
results when this plate is
used is

    (a) clutch capacity will
be increased
    (b) clutch may slip beca
the mean radius is
smaller
    (c) wear rate will be
lower due to the
reduction in slip
    (d) torque transmitted
will be greater since
area has been increa

32. When a body is in equilibrium, the clockwise
moments about any point
equals the anti-clockwise
moments about the same
point. This statement is
known as the

(a) turning moment
(b) definition of balance
(c) principle of moments
(d) definition of the
     centre of gravity

33. Figure 2.26.3 shows a force
    being applied to a lever. The
    turning moment about the
    fulcrum is
    (a)   25 Nm
    (b)   29 Nm
    (c)   100 Nm
    (d)   290 Nm

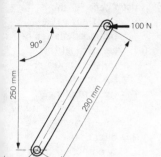

Figure 2.26.3

34. The effective length of a
    drop arm of a steering
    system is 200 mm. When
    the steering box applies a
    turning moment of 170
    Nm to the drop arm, the
    force given to the drag
    link is
    (a)   34 N
    (b)   85 N
    (c)   850 N
    (d)   1176 N

35. A couple is formed by two
    forces of 25 N each which
    act at a distance of 120 mm.
    The magnitude of this
    couple is
    (a)   3 Nm
    (b)   6 Nm
    (c)   25 Nm
    (d)   50 Nm

36. The theoretical point at
    which the total mass of a
    body is considered to act
    is called the
    (a) principle of moments
    (b) turning moment
    (c) counter-balance
         point
    (d) centre of gravity

37. The height of the centre of
    gravity affects the
    (a) angle of roll when
         cornering

(b) load on the front
     and rear wheels
(c) load on the front
     wheels when the
     vehicle is stationary
(d) extra load imposed
     on the rear wheels
     when braking

38. Rearranging the load on a
    vehicle causes the centre of
    gravity to shift towards the
    rear axle. This alteration in
    position causes
    (a) the load on the rear
         axle to decrease
    (b) the load on the front
         axle to increase
    (c) an increase in the
         braking torque
         needed to lock the
         rear wheels
    (d) a decrease in the
         braking torque
         needed to lock the
         rear wheels

39. A rotating shaft is in a
    state of balance when the
    centre of gravity is situated
    on the
    (a) circumference
    (b) axis of rotation
    (c) heaviest side
    (d) lightest side

40. A wheel and tyre is stated
    to be 'out-of-balance'. This
    is normally corrected by
    (a) reducing the mass
         on the heavy side
    (b) moving the centre
         of gravity towards
         the heavy side
    (c) fitting a counter-
         balance mass dia-
         metrically opposite
         the heaviest point
    (d) attaching a weight to
         the rim diametrically
         opposite the lightest
         point

41. When an object moves in a circular path a force acts towards the centre of rotation. This force is called
    (a) centrifugal
    (b) centripetal
    (c) inertia
    (d) reactive

42. A road wheel is mounted horizontally on a balancing machine and rotated at a high speed. During this operation a stone, which was embedded in the tread, is released. Reference to Figure 2.26.4 shows that the path of the stone is indicated by the letter
    (a) A
    (b) B
    (c) C
    (d) D

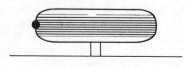

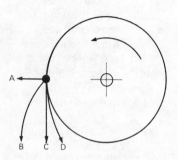

Figure 2.26.4

43. When the centrifugal force acting on one side of the axis of a road wheel is not equal to the force which acts on the other side, then the effect is
    (a) vibration
    (b) slip
    (c) reduced adhesion
    (d) wheel cannot be moved

44. After efficiently balancing a road wheel statically, the wheel is found to be out-of-balance when it is rotated. This is because the
    (a) static balance mass is too small
    (b) static balance mass is too large
    (c) centrifugal forces are forming a couple
    (d) centre of gravity does not coincide with the axis of rotation

45. A four-cylinder in-line engine crankshaft has the crank throws arranged in a manner which gives a firing order of 1 3 4 2 or 1 2 4 3. This arrangement is used because the
    (a) shaft has to be balanced statically
    (b) centrifugal forces must be counter-balanced
    (c) two unbalanced couples can be made to act in the same direction
    (d) two couples must act against each other to obtain satisfactory engine balance

46. One side of a vehicle is raised to a point where it is on the verge of over-turning. The vehicle will overturn when the
    (a) vertical line through the centre of gravity falls on the outside of the wheels
    (b) anti-clockwise moment about the centre of gravity are greater than the clockwise moments
    (c) line from the highest point of the vehicle falls outside the wheelbase
    (d) inward acting forces form a larger couple than that produced by the outward acting forces

47. As applied to machines, the expression

$$\frac{\text{distance moved by effort}}{\text{distance moved by load}}$$

is called the
(a)   movement ratio
(b)   force ratio
(c)   torque ratio
(d)   efficiency

48. As applied to machines the expression

$$\frac{load}{effort}$$

is called the
(a)   movement ratio
(b)   force ratio
(c)   torque ratio
(d)   efficiency

49. As applied to machines the expression

$$\frac{force\ ratio}{movement\ ratio}$$

is called the
(a)   movement ratio
(b)   force ratio
(c)   torque ratio
(d)   efficiency

50. A conventional gearbox transmits the drive through two sets of gears; each set having a ratio of 3 : 1. Assuming friction is neglected, the torque ratio for this gear layout is
(a)   1/9
(b)   1/6
(c)   6
(d)   9

51. A gearbox arrangement consists of two sets of reduction gears; the first set has a pinion of 20 teeth driving a gear wheel of 80 teeth and the second set has a pinion of 40 teeth driving a gearwheel of 60

teeth. If the efficiency is 50 per cent, an input speed of 3600 rev/min will give an output speed of
(a)   300
(b)   328
(c)   600
(d)   655

52. A gearbox has a torque ratio of 2·4 when the gear ratio is 3 : 1. The efficiency is
(a)   1·25 per cent
(b)   72 per cent
(c)   80 per cent
(d)   125 per cent

53. The input and output torque of a gearbox is 70 Nm and 210 Nm respectively. If the efficiency is 75 per cent the gear ratio is
(a)   2·25 : 1
(b)   3 : 1
(c)   4 : 1
(d)   22·5 : 1

54. The two rear axles of a rigid six-wheeled vehicle has semi-elliptical springs mounted to the frame in a manner similar to the Hotchkiss system used on cars. When applied to a heavy vehicle this system has the disadvantage
(a)   load is not divided equally between axles
(b)   torque reaction of each axle cannot be absorbed
(c)   driving thrust from each axle cannot be transmitted
(d)   driving torque tends to lift one axle

55. Although correctly tensioned, a worn vee belt, which contacts the sides and bottom of the pulley, does not transmit full power because the belt
    (a) has lost its wedging action
    (b) has a reduced co-efficient of friction
    (c) surface has become glazed
    (d) has stretched

56. A simple epicyclic gear train gives the greater reduction when the
    (a) sun wheel is driving and the annulus is held
    (b) annulus is driving and the sun wheel is held
    (c) arm is driving and the annulus is held
    (d) sun wheel is driving and the arm is held

57. A ratio of 1·5 : 1 is given by a simple epicyclic gear having an annulus of 100 teeth and a sun of 50 teeth. To obtain this ratio a brake is applied to the
    (a) sun
    (b) planet
    (c) arm
    (d) annulus

58. A sun gear wheel having 60 teeth is held and a planet pinion of 20 teeth is made to revolve around the circumference of the sun wheel. The number of revolutions, relative to an observer, made by the planet to complete one orbit of the sun gear is
    (a) 2
    (b) 3
    (c) 4
    (d) 6

59. A master cylinder of a hydraulic machine has a piston of diameter 20 mm and a ram of diameter 400 mm. If the efficiency is 80 per cent, an effort of 50 N will support
    (a) 0·8 kN
    (b) 1 kN
    (c) 16 kN
    (d) 20 kN

60. A master cylinder piston o diameter 20 mm acts on a ram of diameter 400 mm. To move the ram piston 4 mm requires a master cylinder piston movement of
    (a) 20 mm
    (b) 80 mm
    (c) 400 mm
    (d) 1600 mm

61. Fracture of a rubber pipe connecting a left-hand-front brake to a simple hydraulic system causes
    (a) complete failure of the hydraulic system
    (b) failure of the front brakes only
    (c) the rear brakes to apply only when the pedal is pumped
    (d) failure on the left-h; front brake only

62. The front brake wheel cylinder of a hydraulic system can be made to deliver a greater thrust than that applied at the rear by making the front
    (a) wheel cylinders smaller than the rea
    (b) wheel cylinders larger than the rear
    (c) pipes smaller than the rear
    (d) pipes larger than the rear

63. The wheel cylinders and master cylinder of a hydraulic brake system all have pistons of the same diameter such that a movement of 8 mm of the master cylinder piston caused each wheel cylinder piston to move 1 mm. If four pistons become seized, what is the effect on each of the other pistons?
    (a) The thrust will be doubled and the movement will be 1 mm
    (b) The thrust will be doubled and the movement will be 2 mm
    (c) The same thrust will be applied and the movement will be 1 mm
    (d) The same thrust will be applied and the movement will be 2 mm

64. The velocity of a body is the
    (a) speed
    (b) change in acceleration
    (c) rate at which it changes its speed
    (d) rate at which it moves in a given direction

65. The acceleration of a body is the
    (a) rate at which it moves
    (b) rate of change in velocity
    (c) distance moved in a given time
    (d) distance moved in a given direction

66. Which one of the following represents the approximate value and correct unit for 'acceleration due to gravity'?
    (a) 10 m/s
    (b) 10 m/s^2
    (c) 32 m/s
    (d) 32 m/s^2

67. A vehicle accelerates at a constant rate from rest to a velocity of 48 m/s in 24 seconds. Its acceleration is
    (a) 0·5 m/s
    (b) 2   m/s
    (c) 0·5 m/s^2
    (d) 2   m/s^2

68. A brake efficiency of 100 per cent means that the
    (a) vehicle is stopped instantly
    (b) brake mechanism has no friction
    (c) brakes apply a retarding force equal to the total weight of the vehicle
    (d) coefficient of friction between the brake shoe and drum has a value of 1·0

69. The 'kinetic energy' of a body is the
    (a) energy of motion
    (b) heat produced by the brakes
    (c) work lost to friction
    (d) retarding action given by a force

70. How does the kinetic energy of a vehicle travelling at 100 m/s compare with that given at a speed of 50 m/s? The kinetic energy at 100 m/s is
    (a) less
    (b) equal
    (c) twice as great
    (d) four times as great

2 SCIENCE (d) Materials and materials joining

2.30 Heat treatment

Steel

The main difference between iron and steel is the amount of carbon contained in the composition of the metal. Up to about 1·5 per cent of carbon can be combined with pure iron (or ferrite, which is its chemical name) but above this percentage the carbon remains in an uncombined or free state. The point at which this free carbon or graphite appears is the dividing line between steel and iron — less than about 1·5 per cent the material is steel, and above that point it becomes cast iron. A typical cast iron has a carbon composition of about 3 or 4 per cent, so a large amount of free carbon exists in cast iron.

Carbon has a most important influence on steel: as the carbon content is increased the metal becomes harder and tougher. When the carbon percentage exceeds about 0·3 per cent it is possible to alter the mechanical properties of the steel by heat treatment.

For classification purposes it is possible to use the carbon content to divide steel into three main groups

Low carbon steel	0·05—0·25 per cent
Medium carbon steel	0·25—0·55 per cent
High carbon steel	0·55—0·9 per cent

Other impurities are present in steel, in addition to carbon, namely silicon, manganese, sulphur and phosphorus. To obtain a good quality steel, the amount of each impurity contained in a steel must be controlled.

Critical points

When heat is applied at a constant rate to steel, it is noticed that there are one or more periods when the temperature does not rise. These 'temperature pauses' are called critical points, and during these periods the internal structure of the material is changed. A similar effect takes place when a steel is cooled; if gradual cooling is provided whilst the metal passes through the critical range, the steel will return to its original pre-heated state.

By quenching a medium or high carbon steel from a temperature higher than the critical range, the structural changes will be prevented and increased hardness will result. In all heat treatment operations the important factors are

1. temperature of quenching
2. speed of quenching, i.e. rate of cooling.

The heat treatment processes applied to medium and high carbon steel are: hardening, tempering, annealing, and normalizing.

Hardening

This treatment is performed to give the steel the ability to withstand scratching, wear, abrasion or indentation by harder objects.

Method: Heat the steel to a temperature above the critical range and quench. For general purposes the quenching temperature is estimated by the colour. In the case of hardening, the steel is heated to a 'cherry red'; this represents a temperature of about 800 °C.

The liquid used for quenching is either water or oil; the latter gives a slower rate of cooling and reduces the risk of cracking.

Tempering

Hardening a steel makes it very brittle and destroys its resistance to impact or shock. To improve these qualities, it is necessary to sacrifice hardness by tempering the steel; the final properties are governed by the temperature at which the process is performed.

Method: Heat the steel to the tempering temperature and quench. The tempering temperature is set to suit the conditions to which the steel will be subjected (Table 6). Tempering at 200 °C considerably reduces the brittleness, and tempering at 300 °C decreases the hardness a large extent. Most hand tools are tempered within 200 °C– 300 °C range.

Estimation of the temperature can be obtained by the colour of the oxide film which forms on a polished steel surface.

Table 6. Tempering temperatures

Tool	Temperature (°C)	Temper colour
Scrapers	220	Pale straw
Taps, dies	240	Dark straw
Twist drills	260	Brownish purple
Cold chisels	280	Dark purple
Springs	300	Blue

Some medium carbon steels used for MV components are toughened by tempering at 600 °C. Any excessive heat applied by a mechanic to these components makes the material brittle and weak.

Annealing

The purpose of annealing is to soften the steel and increase its ductility; it also relieves internal stresses.

Method: Heat the steel to a 'cherry red' and cool as slow as possible. Slow cooling is achieved by covering the steel with sand, ashes or lime.

Normalizing

The object of this process is to restore the grain structure of a steel to a strong form after it has been either hot or cold worked. When steel is kept at a red-hot state for a long period of time (e.g. welding), the grain becomes large and coarse. Also when it is cold-worked (e.g. bent without heating), the internal structure becomes deformed and stressed.

Method: Heat the steel to a 'cherry red' and allow it to cool freely in air.

Case hardening

Components such as gudgeon pins and camshafts demand a hard surface to resist wear and a tough core to absorb shocks, so these and many other components are case hardened.

Low carbon steel is used since this ductile material can be made to absorb carbon when it is heated in contact with a carbon-rich substance. The extra carbon absorbed into the steel allows the surfa to be hardened by the normal hardening process; the carburizing time governs the depth of hardness. For lightly loaded parts the depth is only about 0.25 mm, but in cases where large wear is expected, the depth is about 1·5 mm.

The methods used for carburizing are

1. Box process. Steel parts are packed in a box containing a carbon rich substance such as charcoal or bone dust, and heated to about 900 °C for a period of 3 or 4 hours.

2. Cyanide process. Parts are immersed in molten sodium cyanic

3. Open-hearth. Parts are heated to red-heat and dipped into a special compound. By reheating and dipping three or four times, a depth of about 0·1 mm is obtained.

After carburizing, the part should be

1. allowed to cool slowly to anneal

2. heated to a 'bright cherry red' and quenched in oil to refine the grain

3. heated to a 'dull cherry red' and quenched in oil or water to harden the surface.

Most case hardened parts use a steel which contains a small amo of nickel. This material gives a more gradual change of hardness between the skin and core and reduces the risk of the hardened ski 'flaking-off' when in service. *Nitriding* is a low temperature case hardening process used on a special alloy steel and gives an extrem hard case. The machine finished articles are packed, for a period of up to 90 hours, into a tank which is supplied with ammonia gas at temperature of 500 °C. Since a low temperature is used and quenc is unnecessary, distortion is minimised.

Hot and cold working of steel

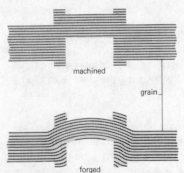

Figure 2.30.1 Grain flow in a crankshaft

Cold working. Cold working is conducted below a temperature of about 600 °C (dull red heat).

When steel is bent or worked the crystal structure is distorted and this makes the steel harder and more brittle. Assuming the material has not cracked, it can be restored to a serviceable condition by annealing.

Hot working. Hot working occurs above the critical temperature range.

Shaping a component by forging is an example of hot working. Forging produces a good grain flow by the shaping and a fine grair by the hammering, so components made in this manner are very strong (Figure 2.30.1).

2.31 Alloying elements

The materials used for motor vehicle construction must meet man requirements, including the following: strength, elasticity, ductilit malleability, brittleness, toughness or shock resistance, hardness, relative density (weight), conductivity (heat and electrical), and c

Some basic materials are able to satisfy a particular requiremen but where this is not possible other elements are often added and alloy is formed.

Table 7 shows the effect of adding an alloying element to a low or medium carbon steel. In practice, alloy steels contain more than one element — a typical composition of one steel suitable for an engine crankshaft is: carbon 0·35—0·44 per cent, silicon 0·10—0·35 per cent, manganese 0·50—0·80 per cent, chromium 0·90—1·20 per cent, molybdenum 0·20—0·35 per cent, sulphur and phosphorus 0·05 per cent maximum.

Table 7. Alloying elements added to low or medium carbon steel

Element	Approximate composition (per cent)	Properties which are improved	Typical application
Nickel	1—5	Toughness, elasticity	Steering and suspension members
	3—5	Case hardening — resistance to 'flaking'	Ball and roller bearing races
Chromium	12—20	Resistance to corrosion (stainless steel)	'Long-life' exhaust systems
Nickel + Chromium	4 } 1	Hardness, ductility, elasticity (can be hardened and tempered)	Shafts High tensile bolts
Vanadium	0·25	Elasticity, toughness, fatigue resistance	Springs, spanners
Tungsten	15	Hardness at high temperature, (high speed steel)	Metal cutting tools
Manganese	12	Strength, hardness and toughness	Axles, starter gears
Molybdenum	0·2—1	Hardness at high temperature, strength, does not become brittle during continuous heating	Crankshafts, gears
Aluminium + Chrome and Molybdenum	1	Hardening ability — allows steel to be case hardened by nitriding process (nitralloy)	Camshafts, crankshafts

.32 Non-ferrous ~~etals~~

Many non-ferrous materials are used in motor vehicles. Some metals such as copper and lead are used in the pure state, but in most cases the properties are improved by alloying a number of metals.

Table 8 shows a summary of the main properties of some non-ferrous metals.

~~n~~nealing of copper

Bending or hammering of copper causes it to become hard and brittle, i.e. it *work hardens*. Also copper has the property of *age hardening*, i.e. it becomes harder with time. To restore the material to its ductile and soft state, it is heated to a dull red heat (650°C) and quenched in water.

~~l~~loys of non-ferrous metals

~~A~~luminium alloy

Pure aluminium is rarely used because of its softness and low strength, but these drawbacks are overcome when it is alloyed with other metals. Many light alloys use aluminium as a base; Table 9 shows the main elements used in two of these.

Table 8. Properties and uses of non-ferrous metals

Metal	Chemical symbol	Melting point (°C)	Main properties	Uses
Aluminium	Al	657	Very light, soft, ductile malleable, good resistance to corrosion, good conductor of heat and electricity	Rarely used in pure state
Copper	Cu	1088	Soft, ductile, malleable, good conductor of heat and electricity	Electrical cables Fuel and oil pipe
Tin	Sn	232	Ductile and malleable	Coating for steel sheets (tin plate)
Lead	Pb	327	Soft, plastic, malleable almost non-elastic, unaffected by most acids	Battery plates
Zinc	Zn	419	Ductile and malleable Non-corrosive in air	Coating for steel (galvanised sheet)

Table 9. Aluminium alloys

Metal	Composition	Main properties	Uses
Duralumin	Aluminium, copper, manganese and magnesium	Good tensile strength	Engine parts which demand strength and lightness
'Y' alloy	Aluminium, copper, nickel, magnesium and silicon	Light, strong, good heat conduction	Pistons, cylinder heads

Table 10. Copper alloys

Metal	Composition	Main properties	Uses
Brass	Copper Zinc	Great ductility, good strength, resistance to corrosion	Radiator parts, lamp fittings, chromium plate parts, light duty bushes, nuts
Bronze	Copper Tin (Phosphor bronze includes about 1 per cent phosphorus)	Good bearing material — resistance to abrasion and low friction qualities	Plain bearings and bushes. Worm-wheels
Copper—lead	Copper Lead	High duty bearing material	Big end and main bearings
Lead—bronze	Copper Tin and lead	Heavy duty bearing material	Heavily loaded engine bearings

Some light alloys have the property of *age-hardening*; in the case of duralumin, a soft, workable condition is only retained for a few hours after heat treatment. Annealing is performed by heating to about 375 °C and allowing the metal to cool in air, water or oil. (Heating by a flame causes ordinary engine oil applied to a metal surface to turn black at about 375 °C so this may be used to indicate the temperature.)

opper alloys

Table 10 shows properties of various copper alloys.

in alloys

Solder and white metal bearing alloy are examples of tin alloys. Solder consists of tin and lead and the proportions used of these are governed by the application.

earing alloy

In the past, the white metal used for crankshaft bearings was Babbitt metal; this had a composition of tin, copper and antimony. Modern engines having high bearing loads demand a low friction material which is strong and resistant to the fatigue caused by surface deflection. Tin alloyed with aluminium is one modern material which combines strength with softness.

inc alloys

Many motor vehicle parts such as carburettors, hydraulic pump bodies, decorative fittings, are die cast, since this method produces a casting which has an excellent finish. The material used is normally a zinc based alloy called Mazac — the name gives an indication of the composition — magnesium, aluminium, zinc and copper. Oxidation occurs with age, and the brittleness which results means that components made in this material must be treated with care.

Examples of non-metallic materials are given in Volume 1 of this book and many other applications of metals are stated in *Fundamentals of Motor Vehicle Technology*.

.33 Nature of stress

Earlier work showed that a load applied to a component caused the particles of the material to be deformed. The extent of this deformation is indicated by considering the load which acts on a given area. These two factors give the stress in a material, so

$$\text{stress} = \frac{\text{load}}{\text{area}}$$

If the load is in newtons and the area in square metres then

$$\text{stress} = N/m^2$$

Other units used for stress are: N/mm^2, pascal and bar.
For conversion purposes

$1\ N/mm^2 = 10^6\ N/m^2 = 1\ 000\ 000\ N/m^2$ OR $1\ MN/m^2$
$1\ Pa\qquad = 1\ N/m^2$
$1\ bar\qquad = 10^5\ N/m^2$ OR $100\ 000\ N/m^2$

Direction of the loading indicates the type of stress; the main stresses are tensile, compressive and shear.

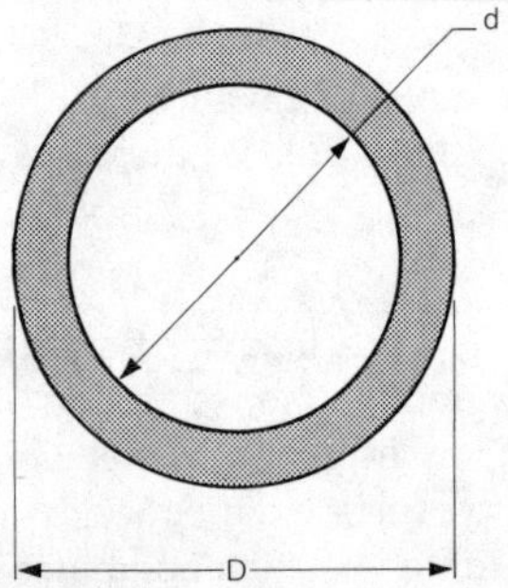

Example 1

A hand brake cable has a cross sectional area of 7 mm^2. Calculate t
tensile stress in the cable when it is subjected to a force of 560 N.

$$\text{Stress} = \frac{\text{load}}{\text{area}}$$

$$= \frac{560}{7} = 80 \text{ N/mm}^2$$

$$= 80\ 000\ 000 \text{ N/m}^2 = 80 \text{ MN/m}^2$$

Figure 2.33.1

Example 2

A hollow gudgeon pin of external diameter 16 mm and internal
diameter 12 mm is subjected to a load of 5·28 kN (Figure 2.33.1).
Calculate the shear stress in the material. In this case the pin is in
double shear since both sides of the pin are resisting the load. So

$$\text{Cross sectional area of pin} = \frac{\pi D^2}{4} - \frac{\pi d^2}{4}$$

$$= \frac{\pi}{4}\ (D^2 - d^2)$$

$$= \frac{22}{7 \times 4}\ (256 - 144)$$

$$= \frac{22}{28} \times 112$$

$$= 88 \text{ mm}^2$$

$$\text{stress} = \frac{\text{load}}{\text{area resisting shear}}$$

$$= \frac{5280}{2 \times 88} \text{ N/mm}^2$$

$$= \frac{5280}{176}$$

$$= 30 \text{ N/mm}^2 \text{ OR } 30 \text{ MN/m}^2$$

Example 3

The diameter of the wire in a suspension spring is 15·478 mm, bu
due to corrosion this diameter is reduced to 12·7 mm. What will b
the effect of this corrosion on the
(a) stress in the spring
(b) altitude or 'riding height' of the vehicle

(a) Stress is equal to load/area, so the effect of the corrosion i
the area will be reduced and this will increase the stress.
(b) Hooke's law shows that the deflection of a spring is propo
to the stress. An increased stress in a suspension spring will cause
'riding height' of the vehicle to be decreased, i.e. the vehicle body
be nearer the ground.

xcessive loading on materials
id components

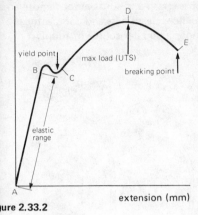

ure 2.33.2

When a metal is subjected to a tensile load the material extends. Figure 2.33.2 shows the result obtained from a test on a ductile mild steel specimen. In common with other metals, the extension of this steel varies with the load. Important parts of this graph are

    1. A-B: Extension is proportional to the load, i.e. Hooke's law applies. Provided the load does not exceed point B, the material will return to its original length when the load is removed. The working stress in a component should always be within this elastic range.

    2. B-C: Material yields and results in the material extending without further increase in the load. Once this point has been reached the plastic state of the metal produces a permanent deformation. (The yield of a metal can be felt when overtightening a bolt — when the yield point is reached, the bolt's resistance to rotation decreases and this causes a considerable reduction in its clamping ability.)

    3. D: Maximum load (or ultimate tensile stress) taken by a material. Beyond this point the metal starts to break.

Consideration of the behaviour of a metal when subjected to a stress shows that overloading can cause: breakage and the loss of elastic properties. To avoid these effects the material is given a *factor of safety.* This factor ensures that the working stress is a fraction of the metal's ultimate tensile stress. Tightening a bolt to a recommended torque gives a safe working stress and so reduces the risk of both under and overloading.

.34 Oxy-gas
elding

The process of fusion welding, which entails the melting together of the materials, uses either oxy-gas or electrical energy to provide the heat to melt the materials. This heat must be intense and localized since the melting of the metal should be confined to the region of the join. When welding body panels, the restriction of the heat to a small area is essential, since any expansion causes distortion which is difficult and costly to rectify. For this reason spot welding proves to be advantageous for this type of work.

    Oxy-gas equipment has many motor vehicle uses. Besides the many welding tasks which can be undertaken, the equipment also provides the mechanic with a source of intense heat which can be applied to a number of jobs. The ease in which the equipment can be used in this manner can lead to dangerous practices unless basic safety precautions are observed. A naked flame should not be used in situations similar to the following unless the area is rendered safe.

    1. Petrol tanks — the vapour remaining in a fuel tank is highly explosive. Even after repeated flushing with water the risk is still great so before attempting fuel tank repairs it is advisable to study the pamphlet, *Repair of drums or tanks which have contained petrol or other inflammable liquid* issued by the Home Office.

    2. Flammable substances — underseal paint, body trim, fuel lines, etc., are a fire hazard when heated.

    3. Plastics — many plastics materials give off poisonous gases when heated.

    4. Confined spaces — poor ventilation in places like workshop pits allow petrol vapour to collect which can produce an explosion.

Oxy-acetylene welding equipment

Oxygen and acetylene are the gases generally used for gas welding (Figure 2.34.1). Normally they are supplied in steel cylinders (bot as they called in the workshop) although some welding plants mak their own acetylene from calcium carbide in a gas generator. Acet made in a portable generator is cheap and suitable for remote area the country, but the convenience, cleanliness and compactness of 'bottled gas' makes the cylinder a popular source of supply.

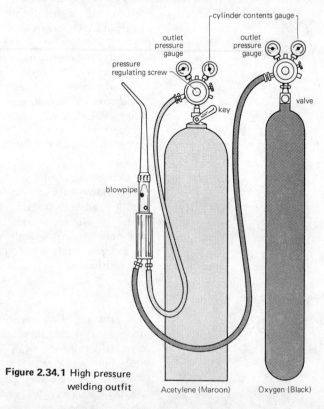

Figure 2.34.1 High pressure welding outfit

Acetylene (Maroon)   Oxygen (Black)

Oxygen cylinder

Oxygen is stored at a pressure of 172 bar (17·2 MPa) in a steel cylinder which is painted *black* for identification purposes. The cylinder is fitted with a shut-off valve and connection to the press regulator is by means of a *right-hand thread.* Oil or grease must n applied to any thread since pure oxygen can cause combustion of oil or grease. Leakage can be traced by using soapy water.

Acetylene cylinder

Acetylene is a gas lighter than air and consists of carbon and hydrogen, as shown by the chemical symbol C_2H_2. It can be ignited by a spark and in air it burns with a sooty flame. With copper, acetylene forms an explosive compound, so copper fittin should not be used.

When acetylene is compressed it is unstable and explosive, so f safety reasons the gas is dissolved in liquid acetone. This liquid is absorbed into a porous substance such as kapoc and contained in steel cylinder painted maroon. Maximum pressure of the acetyler is about 15 bar (1·5 MPa) and various cylinder capacities are avail Cylinders containing combustible gases such as acetylene have *lef hand threads* on all equipment. Since acetylene is highly inflammable, leakage must be avoided and a naked flame must not be used to search for leaks—these can normally be detected by the pungent smell or by applying soapy water.

As the gas is consumed, the pressure in the cylinder lowers; this causes some of the dissolved gas to be released from the acetone thus providing a constant flow of acetylene gas. Provided that the volume of acetylene drawn from the cylinder is not excessive, the gas supplied will not contain more than the minimum amount of acetone. Since the acetylene cylinder contains liquid, the cylinder should be used when it is positioned vertically with the valve at the top.

essure regulator

A reducing valve or regulator is needed for both gases to lower the pressure to 0·6 bar (60 kPa) or less to suit the blowpipe and welding process.

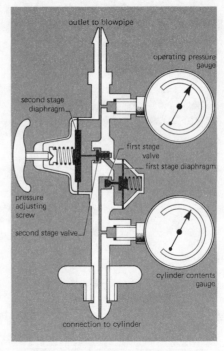

Figure 2.34.2 Two stage pressure regulator

Figure 2.34.2 shows one type of pressure regulator. This reduces the pressure in two stages; the first stage lowers the pressure by a fixed amount and in the second stage an adjusting screw enables the final pressure to be controlled by the operator. Each stage consists of a valve, spring and diaphragm; the spring is positioned so that it forces the valve toward the 'open' position.

A gas flow through the regulator causes the gas pressure to act on the diaphragm, and when this pressure is sufficient to overcome the spring, the valve starts to close. Restriction of the flow by the valve results in a reduced pressure being applied to the diaphragm, so a point will be reached where the diaphragm pressure will balance the spring pressure and a steady flow at a set pressure will be obtained.

The welding blowpipe

Many different types of blowpipe or torch are available; Figure 2.34.3 shows a typical high-pressure type. Two control valves regulate the volume of each gas fed to the mixer and nozzle. A number of nozzles are required if many different thicknesses of metal are to be welded. The orifice of the nozzle can be identified by the number stamped on its side; this number indicates the consumption of each gas per hour.

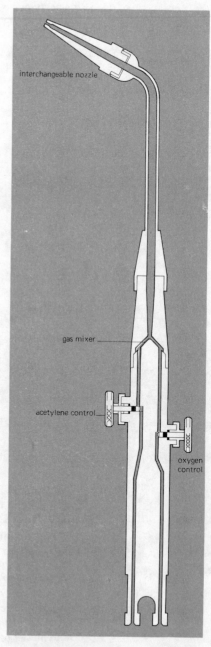

Figure 2.34.3 High-pressure blowpipe

Eye protection

It is essential that the eyes of a welder are protected from the brightness of the welding process and from the spatter of red hot particles which erupt during it. This protection must also be extended to other persons assisting the welder. Special goggles containing coloured lens, which are protected by replaceable plain glass discs, can be obtained in many patterns.

In addition to eye protection it is recommended that protective clothing, which should take the form of a long leather apron, be worn. Without this protection the mechanic, dressed in oily overalls, who undertakes the occasional welding repair is gambling with his life. There have been many instances where welding goggles have prevented the mechanic from seeing that his clothing was burning until it was too late.

Storage and handling of welding equipment

The main safety aspects to be observed are
1. Gas cylinders must be handled with care — do not drop and do not bump the cylinder valves.
2. Gas cylinders must not be heated in any manner. If for any reason an acetylene cylinder becomes overheated then it should be taken outdoors, sprayed continually with water and have its valve opened to release the gas. The supplier should be notified of the incident.
3. Oil or grease must not be used on any part of the equipment.
4. Keep hoses clear of any source of heat.
5. Turn off gas supply when equipment is not being used or when equipment is being moved to another location.
6. Never use a naked flame to test for leaks.
7. Ensure that cylinders are secure and are not likely to fall over.
8. A lighted blowpipe should not be left unattended.
9. When setting up work the flame should be extinguished or directed away from the operator and assistant.
10. Protective clothing and goggles should be worn.
11. Never use cylinders for work supports or rollers.
When setting up welding equipment:
1. Blow out dirt from cylinder outlet by opening and closing the valve before fitting regulator to new cylinder.
2. Do not attempt to fit a regulator to a cylinder having a damaged thread.
3. Open cylinder valve slowly to avoid damaging regulator.
4. After opening cylinder valves leave key in acetylene valve in case of emergency.
When shutting down equipment
1. Do not overtighten cylinder valve but tighten it sufficient to prevent gas leaks.
2. Vent pipes of gas by opening blowpipe valves.
3. Remove load from regulator spring.

Care of the blowpipe

When the blowpipe is not in use the hoses should be coiled-up clear of the ground and the blowpipe retained in a secure position.

Control valves should not be overtightened and on no account must oil or grease be applied to any part. If lubrication is needed then powdered graphite can be used (Note: do NOT use graphite grease).

Nozzle tips should be cleaned periodically. A strand of copper wire can be used as an orifice probe to remove soft deposits, but burrs will require the use of an orifice reamer—these are usually supplied as a set and are invaluable since a partially blocked nozzle gives an irregular, poor flame.

Backfiring into the blowpipe can be caused by the

1. regulator pressure being too low
2. blowpipe touching the metal
3. hot metal adhering to the blowpipe
4. overheating of the blowpipe
5. low velocity of gas at nozzle

If, as the result of a backfire, black smoke discharges from the nozzle and a squealing sound emits from the blowpipe, then the gas supply should be shut out immediately to prevent damage. The fault can be caused by trapped metal particles in the pipe or under the valves. Before relighting, the torch should be dismantled and cleaned.

A flashback, which is the term used when the flame travels back through the hoses, is prevented by fitting a one-way valve, called a flashback arrester, in each hose.

If at any time a flame is given off from a hose or valve, then the acetylene cylinder valve should be turned off immediately.

Structure of oxy-acetylene flame

Having appreciated the potential dangers associated with the equip if handled incorrectly, it is now possible to examine the type of fl used for welding.

The sequence of lighting the flame is

1. Select the nozzle to suit the material being welded and fit to the blowpipe.
2. Open valves on oxygen and acetylene cylinder about ½ turn. Leave key in acetylene valve.
3. Adjust regulators to give required pressure. This adjustment carried out by opening the appropriate valve on the blowpipe and screwing in the regulator control until the desired pressure is obtai Shut off each valve on the blowpipe after the correct setting is achieved—do not overtighten the valves.
4. Turn on the acetylene and ignite the gas. If the flame is smo the valve should be opened further until a smoke-free flame is obt Increasing the gas velocity beyond this correct setting for the nozz causes the flame to move away from the nozzle tip.
5. When the correct amount of acetylene is obtained, open the oxygen valve to the appropriate setting.

As the flow of oxygen increases, the flame changes to give setti which are called: carbonizing, neutral, and oxidizing.

Carbonizing flame

A carbonizing flame is produced by insufficient oxygen or excess acetylene and identified by the 'whitish' feather which appears beyond the inner blue luminous cone at the tip of the nozzle (Fig 2.34.4). As the name suggests, this flame has an excess of carbon, if a carbonizing flame was used for normal mild steel welding, the carbon would be absorbed into the material and this would cause brittleness.

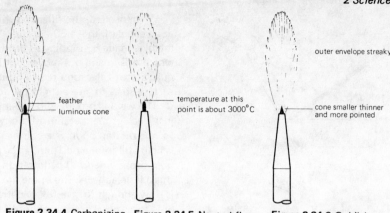

Figure 2.34.4 Carbonizing flame (excess acetylene) **Figure 2.34.5** Neutral flame (used for most welding) **Figure 2.34.6** Oxidizing flame (excess oxygen)

Some types of work, e.g. Lindewelding of large gas and oil pipe lines and hard surfacing of components, require this type of flame.

Neutral flame

As the oxygen supply is increased the length of the 'acetylene feather' decreases. At the point where the feather has just been eliminated a neutral flame is obtained and this is the type which is used for most welding operations (Figure 2.34.5).

The inner cone is a region where the gases are being partially burnt to form carbon monoxide and hydrogen. When these products meet the oxygen from the atmosphere, the final burning takes place and this produces the outer envelope of the flame. Oxygen is taken from the air in the vicinity of the weld to supply the outer envelope, so this means that only a small amount of oxygen is left to combine with the red hot iron, i.e. oxidation of the surface does not occur to any large extent during oxy-acetylene welding.

Oxidizing flame

Opening the oxygen valve beyond the neutral position causes the inner cone to become smaller, thinner and more pointed and the outer envelope to appear streaky. This flame is called oxidizing since it contains an excess of oxygen (Figure 2.34.6). Although slightly hotter than the neutral flame, the oxidizing flame is unsuitable for normal welding due to oxidation problems. However, the flame is used in some instances, e.g. bronze welding, a flux being used in this case to prevent the metal becoming oxidized.

Welding methods

The technique used for oxy-acetylene welding on a motor vehicle is often dictated by the situation, however when welding material of thickness up to about 6 mm, the *leftward* method of welding should be used.

Leftward or forward welding

Assuming the blowpipe is held in the right hand, the weld is started at the right-hand side of the metal and progresses in a leftward direction. A neutral flame, forming an angle of about 60° to the surface, is directed down the join to preheat the metal. Good fusion of the edges is achieved by moving the torch in either a zigzag or circular motion as shown in Figure 2.34.7.

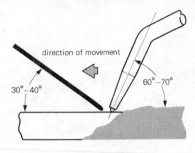

Figure 2.34.7 Leftward welding technique

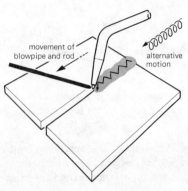

When needed, the filler rod should be held in line with the join at an angle of about 30° to the surface. Application of the filler metal is made by dipping the rod into the molten metal; the rate at which this in-and-out movement is performed is governed by the quantity of filler required to build up the weld just proud of the surface so as to give a slight reinforcement.

The weld is produced by keeping the inner cone of the flame just clear of the metal until the two surfaces fuse together and form a molten pool. The complete action can be shown if the torch is held still in one position; after fusion occurs the metal level in the pool drops when full penetration is obtained, and a short time after this a hole appears. Blowpipe movement should be at a rate which is slow enough to give full penetration but fast enough to prevent a hole forming. At no time should the inner cone touch either the parent metal or the filler rod.

Thickness of metal (mm)	Diameter of welding rod (mm)	Edge preparation
Less than 0.9	1.2-1.6	
0.9-3	1.6-3	0.8-3mm
3-5	3-3.8	80° 1.6-3mm

Figure 2.34.8 Edge preparation

Figure 2.34.8 shows that no bevelling is necessary for material of thickness less than about 3 mm, above this thickness the metal is bevelled to form a 'V'. Using the leftward method for welding mild steel plate thicker than about 6 mm causes the molten metal to be blown along the 'V' in front of the torch. Since this metal adheres to the parent metal instead of fusing with it, a poor weld results. Although the problem can be minimized by making a series of runs a more suitable solution is to use other methods, e.g. the rightward technique.

The features of a good and bad weld are shown in Figure 2.34.9

One of the most important features is the fusion between the weld metal and the parent metal. Many welds appear neat on the surface, but on bending the plate, the weld parts from the parent metal and reveals a very weak joint. This is generally caused when the filler rod is used like a stick of solder, i.e. it is melted into the join to adhere to the surfaces instead of being used to build-up the weld after fusing the two surfaces together.

Poor penetration and formation of a 'valley' down the weld are obvious defects.

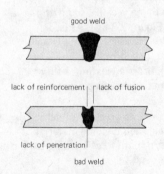

Figure 2.34.9 Bad weld

Fillet welds

Owing to the greater mass of metal at the joint, a larger nozzle is used for fillet welding (Figure 2.34.10) than that used for butt welding. The lap joint, which requires fillet welding, is simpler to weld than the vertical fillet, since undercutting cannot occur. Und

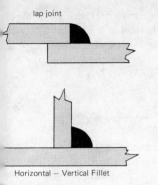

lap joint

Horizontal — Vertical Fillet

Figure 2.34.10 Types of fillet welds

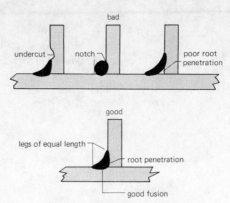

bad

undercut — notch — poor root penetration

good

legs of equal length — root penetration

good fusion

Figure 2.34.11 Fillet weld faults

cutting (Figure 2.34.11) can be avoided by directing the cone of the flame towards the plate which is capable of absorbing most heat and by using torch and rod angles as shown in Figure 2.34.12.

This technique of directing the flame towards the greater mass of metal is also used when joining plates of different thickness.

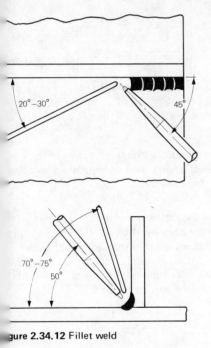

20°–30°   45°

70°–75°
50°

Figure 2.34.12 Fillet weld

hot

cold

before   before

after   after

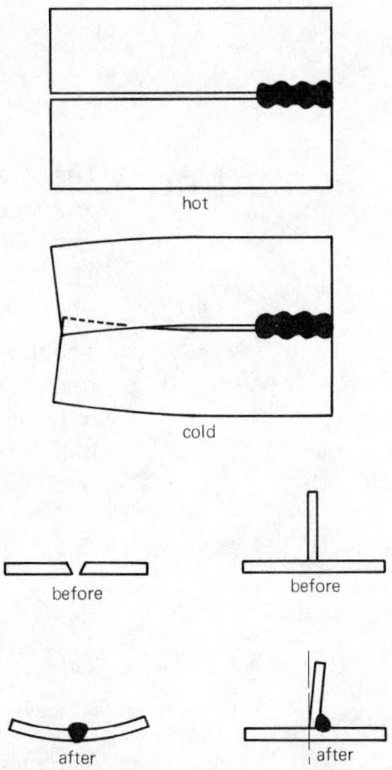

Figure 2.34.13 Distortion caused by welding

Distortion

When metal cools it contracts. Applying this statement to a welded joint shows that distortion will occur unless precautions are taken. Figure 2.34.13 shows the effect of contraction during the welding of two plates: the larger the flame the greater will be the distortion. Performing this test on thicker material results in a smaller amount of distortion — instead, the uneven contraction causes internal stresses to be set up.

Motor Vehicle Technology

To minimize distortion in light gauge materials, the metal should be either tacked at intervals or securely held in a jig (Figure 2.34.14 Where distortion is critical the flame should be as small as possible and heat should be kept from the surrounding metal by insulating material or by sponging with water.

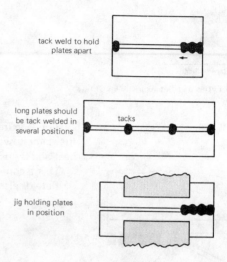

Figure 2.34.14 Methods for minimizing distortion

2.35 Oxy-gas cutting

The ease in which iron or steel is cut with oxy-acetylene, oxy-propane, or similar gases makes this method particularly attractive, especially when thick metal has to be cut.

Cutting action

The cutting takes place in two stages: the metal has to be heated to a bright red and then a high pressure jet of oxygen is directed onto the hot metal. This oxygen causes the iron to oxidize and since oxidized iron melts at a low temperature, the molten metal will be blown away by the gas stream. Once the cut has started much of the heat will be obtained from the cutting process.

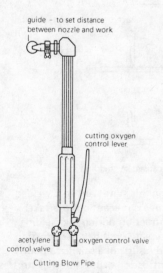

Figure 2.35.1 Oxy-gas cutting blowpipe

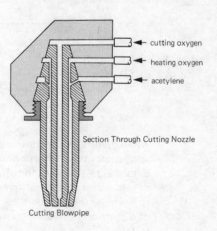

utting blowpipe

Figure 2.35.1 shows a blowpipe used for cutting steel. It comprises an oxy-gas system for the preheating flame and a separate high pressure oxygen supply which is fed to the centre of the nozzle when the lever is operated.

A nozzle guide enables the distance from the nozzle to the work surface to be controlled; for material of thickness up to 50 mm a distance of about 3 mm is required.

se of cutting blowpipe

The acetylene regulator pressure required when using a cutting blow-pipe is similar to that used for welding material of equal thickness to that being cut. Oxygen pressure is much higher and the actual setting is governed by the type and size of nozzle used. For material of thickness 25 mm, a typical pressure is 2 bar (200 kPa).

The heating flame should be set to the neutral form and the torch is applied at a right angle to the edge of the material farthest from the operator. When a bright red heat is obtained, the oxygen lever is pressed and the torch is then drawn slowly towards the operator. If the movement is too fast the cutting action will cease and reheating of the metal at the point where the cut ended is necessary.

Red hot molten metal streaming from a cut creates a fire hazard and can cause severe burns. Safety precautions in respect to these aspects must be taken before the work is started.

.36 Metal rc-welding

At some time, whilst working on a motor vehicle, it is likely that you will have accidentially shorted the 'live' battery lead to earth. Contact between the lead and earth causes a high current flow, and when the surfaces are initially parted this current jumps the air gap. The passage of electricity through air, seen as a spark or electric arc, is possible when the voltage is sufficient to make the air become a conductor by creating ionization. Assuming that the two surfaces in question have not fused together, then separating each part sufficient to give a small air gap will, most probably, only cause an initial flash, but if the voltage is higher than 20 V it is possible for the current to continue to flow. The arc formed has a temperature of about 4000 °C, so metals in the vicinity of the arc will be melted very quickly. As the air gap is increased, the resistance to electrical flow is also increased, so unless the voltage is raised, the arc will cease.

This example is similar in principle to electric arc welding (Figure 2.36.1). The plate to be welded is connected to one electrical supply lead; the other lead is attached to a filler rod which acts as an electrode. After striking the arc by touching the plate with the electrode, a small arc gap of about 5 mm is set. The continuous arc formed will fuse the two plate surfaces together and will melt away the electrode at a suitable rate to provide the filler material.

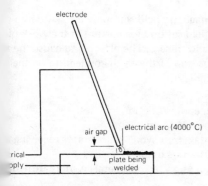

ure 2.36.1 Metal arc-welding

lding equipment and formation

Metal arc welding demands a voltage of 60—90 V for striking the arc, but once the arc is formed a much lower voltage is required (e.g., 25 V) to maintain the arc. Current flow, for a given voltage, will be affected by the diameter of the electrode and the length of arc; a typical value for an electrode of diameter 3 mm is 120 A. To make

the equipment suitable for a wide range of work, it should be possib
to vary the output current.

These requirements are met by modern welding plants. The two
basic types are the direct current welding generator and the alterna
current mains transformer.

Welding generator (d.c.)

Driven by either an electric motor or engine, this generator is simila
to the d.c. dynamo used on a vehicle, except that the output of a
welding generator is much higher. Voltage control is achieved by
varying the current supplied to the field circuit; a socket plug and
rheostat enables the operator to select the appropriate output.

Mains transformer (a.c.)

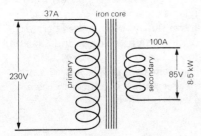

Figure 2.36.2 Transformer action

The lower initial cost of this type makes it an attractive unit for m
vehicle repair establishments. It consists of an oil-cooled mains trai
former, which steps-down the voltage to about 85 V by using a sec
dary coil having a smaller number of turns than the primary (Figur
2.36.2).

Regulation of the output current can be achieved by either var
the strength of the transformer field or by using a choke, which cc
of an iron core around which is wound a coil (Figure 2.36.3). Welc
current is altered by
1. a selector switch, which taps the choke coil at different poir
2. moving the iron core in or out of the choke coil
These methods of output control of a.c. equipment enables the
arc to be struck with the ease associated with d.c. units.

The chance of an operator receiving a fatal electric shock is pre
if a fault develops in the transformer, so to avoid this risk, the tra
former and work-piece should be efficiently earthed.

Compared with d.c. equipment, the a.c. welding transformer h
the following advantages
1. Lower initial cost
2. More portable
3. Less maintenance
4. No arc-blow due to magnetic field set up by arc. (Arc-blow
occurs when the magnetic field set up by a current in the work-pi
interacts with the field around the arc. The effect is to cause the a
to wander. Changing the position of the earth connection on the
piece may overcome the problem.)

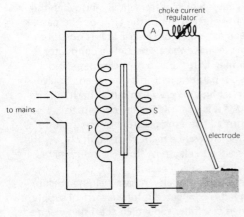

Figure 2.36.3 A.C. welding transformer

The disadvantages of a.c. equipment are
1. Covered electrodes must be used
2. Greater care must be taken to avoid the operator receiving a
electric shock as a higher voltage is used
3. Welding of cast iron and non-ferrous metals is more difficul

Electrode polarity

The arc is formed by negatively charged particles called electrons
which flow from the negative pole through the ionized air to the
positive pole. Energy released by the arc appears as light and hea
Electron flow in one direction, as given by d.c. equipment, cause
2/3 or the total heat from the arc to be generated at the positive
pole, the remaining 1/3 being developed at the negative pole (Fig

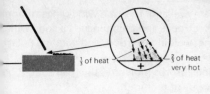

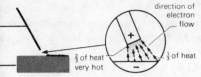

Figure 2.36.4 D.C. — electrode polarity

2.36.4). Arranging the polarity so that the electrode rod forms the positive pole, causes the rod to burn away at twice the rate produced by a rod of negative polarity. This feature can be used to prevent rapid burning of a rod, e.g. bare rods and medium coated rods should form the negative pole, whereas a heavily coated rod is best suited to the extra heat obtained from the positive pole.

As the term suggests, alternating current flows in one direction and then the other. As applied to the mains (in the U.K.) this change occurs fifty times per second, i.e. the frequency is fifty cycles per second. Since a.c. equipment produces an alternating electron flow, the heat developed at each pole is the same (Figure 2.36.5).

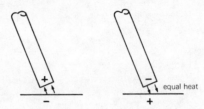

Figure 2.36.5 A.C. — electron flow changes many times per second so heat at each pole is equal

Effect of voltage on d.c. welding

Voltage required to strike an arc with d.c. equipment is about 60 V (80–90 V with a.c. systems). Once the arc has been struck, the voltage drop across the arc should be

1. 15–20 V for bare wire, or lightly coated rods using a short arc
2. 20–25 V for medium coated rods and longer arc

The voltage drop governs the heat generated at the arc, so if a large amount of penetration is needed a high voltage is necessary. Normally the limits of this voltage are 14 V and 40 V; below 14 V the arc cannot be maintained and above 40 V a poor weld is obtained.

Besides controlling the voltage at the generator, it is also possible to increase arc voltage by

1. coating the rod with a substance which feeds the arc with hydrogen — this gas has a lower electrical conductivity than air
2. increasing the length of the arc. A long arc is not normally recommended because it is difficult to control. In addition, it causes the large area of molten metal under the arc to absorb oxygen and nitrogen when it becomes exposed to the air; this results in a poor weld

Electrode coatings

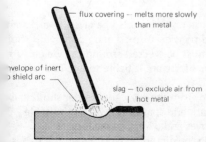

Figure 2.36.6 Action of flux coating of rod

When a bare wire is used as an electrode rod, it is found that

1. the arc is difficult to control; it tends to wander from the spot where the heat should be directed and often results in poor fusion of the work piece
2. the weld is porous, brittle and weak due to absorption of oxygen and nitrogen
3. slag is contained in the weld

To reduce these defects the electrode rod is normally coated with a suitable flux. This melts to form an envelope of inert gas to shield the arc and molten metal; afterwards it solidifies to exclude the air from the weld (Figure 2.36.6).

Many different coatings are available to suit a wide range of appl
cations. BS 1719 classifies electrodes and recommends a code which
indicates the type of covering, welding position, and welding polari
and current.

Since the electrode coating absorbs moisture, the rods should be
kept in the box in a dry place until required.

Electrode holder

The electrode holder consists of a quick acting clamp made of copp
and a handle which is insulated to protect the operator from heat a
electric shock (Figure 2.36.7).

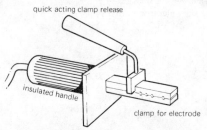

quick acting clamp release

insulated handle

clamp for electrode

Figure 2.36.7 Electrode holder

Face shield and protective clothing

An electric arc produces an intense light which contains infra-red
and ultra-violet rays. These rays are very harmful to the eyes and i
directed onto the skin can cause serious burning. The face shield
should be of an approved pattern and Figure 2.36.8 shows two bas
types. Both masks use a special lens (darker than that used for gas
welding) and this is protected on both sides by plain glass which is
cheaper to replace when the surface becomes damaged.

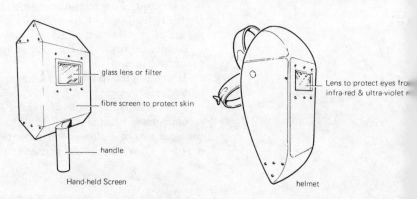

glass lens or filter

fibre screen to protect skin

handle

Hand-held Screen

Lens to protect eyes fro
infra-red & ultra-violet r

helmet

Figure 2.36.8 Head shields

Protection from the intense rays must also be given to assistant
and, in cases where the welding is performed in an open workshop
screen or fireproof curtain must be erected around the welder. An
person receiving the full rays of the arc is liable to eye injury, e.g.
'arc-eye'.

Besides screening the skin from the rays, the welder should also
protect himself from the spark and spatter by wearing suitable
clothing such as asbestos gloves and a long leather apron.

General purpose clear goggles should be worn when the slag is
chipped off the metal after welding.

Equipment manipulation

Thickness of metal (mm)	Electrode size (mm)	Plate preparation
Less than 3	2	
3–6	2·5–4	1·5 3mm
6–20	4–5	60° 3–4mm more than one run is necessary for plates thicker than 8 mm

Figure 2.36.9 Plate preparation

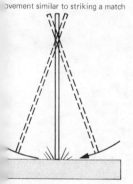

movement similar to striking a match

Figure 2.36.10 Striking the arc

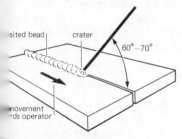

deposited bead | crater | 60°–70°

movement towards operator

Figure 2.36.11 Downhand method

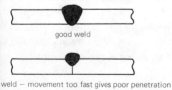

good weld

weld – movement too fast gives poor penetration

weld – movement too slow gives undercutting & excessive metal deposit

Figure 2.36.12 Effect of speed of movement

Let us consider the method used when two pieces of mild steel are to be butt welded by using the 'down-hand' technique.

1. Set up the metal so that the butted edges or gap between the plates runs in a direction towards the operator. Plate preparation is shown in Figure 2.36.9.

2. Select the electrode to suit the material to be welded. The packet in which the rods are supplied normally specifies the current setting and also the electrode polarity for d.c. equipment (see Table 11).

The effect when the current is:
(a) too low, is poor penetration and small amount of metal deposited,
(b) too high, is undercutting and excess spatter. (Spatter is the name given to the metal globules which erupt from the molten pool and form along the edge of the weld.)

Table 11. Typical current values for general purpose electrode* used for welding mild steel.

Electrode diameter (mm)	Welding current (A)
2	40–70
2·5	75–100
3·25	95–125
4	135–180
5	155–230

*When used with d.c. equipment the electrode should be connected to the positive pole.

3. Connect the return clamp securely to the workpiece. After ensuring that the electrode is clear of all earth points, 'switch-on' the equipment.

4. Strike the arc by touching the electrode on the workpiece at the point farthest from the operator. By moving the rod in a manner similar to that used when striking a match, the tendency for the rod to 'stick' to the plate is reduced (Figure 2.36.10). As the rod is moved to strike the arc, the face shield is set to protect the eyes. On no account should the arc be struck before the eyes are guarded from the intense light.

5. With an arc length of 3·5 mm and the rod forming an angle of 60–70° with the workpiece, the rod is gradually moved towards the operator (Figure 2.36.11). The rate of movement should be that which gives an even, regular pattern with no holes. Incorrect speed, as shown in Figure 2.36.12, causes the following effects:

(a) too fast – poor penetration and insufficient metal deposit
(b) too slow – excessive metal deposit, crater too deep and electrode becomes red hot

Rods having a heavy flux coating produce a thick slag deposit which, when molten, appears lighter in colour than the molten metal. This slag should not be allowed to form in front of the weld since it will result in blow-holes.

If the weld has to be stopped at any mid-point for any reason, e.g. fitting a new rod, then the slag must be chipped from the end of the bead and removed from the crater by wire-brushing. When the weld is restarted the arc is struck at the forward end of the crater and the rod is then moved back to continue the regular weld pattern.

Butt welding of material thicker than 8 mm requires a number of runs in order to build up the metal to the required thickness. Complete deslagging is necessary between each run.

Lap and fillet welding

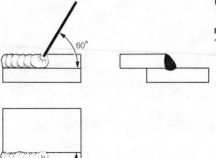

The electrode angles for a lap joint welding are shown in Figure 2.36.13. No preparation is necessary since the edge of the metal form a 'V'.

Fillet welding is more difficult (Figure 2.36.14) because the metal must penetrate to the corner of the plates without causing undercut. To achieve this, a short arc and slow movement is required.

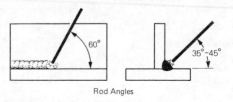

Rod Angles

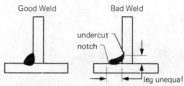

Figure 2.36.13 Lap welding

Figure 2.36.14 Fillet weld

Welding and semi-conductors

Damage will occur to semi-conductor devices if they are either over heated or subjected to induced currents such as those produced when electric welding is performed on a vehicle.

As a precaution against damage

1. no welding should be performed in the vicinity of a semi-conductor device

2. before electric welding is carried out on a vehicle, the unit containing the semi-conductors should be disconnected from the main electrical circuit.

Shielded arc welding process

The shielded arc method enables materials such as aluminium, stainless steel, copper and steel to be fusion welded. The arc and molten metal are shielded by an envelope of inert gas to prevent the atmosphere from forming oxides and nitrides, which cause weakness and brittleness. Although the flux coated electrode does a similar job, the molten slag produced by the flux can become trapped in the molten metal and this gives a weak weld. Also, the removal of the hard slag, formed on the surface of the weld after the metal has cooled, creates an additional, irksome task. These problems may be overcome by using an inert gas, such as argon or carbon dioxide, which is supplied from a gas cylinder.

Argon-arc process

There are two basic types of equipment in use; these are: TIG — tungsten, inert gas, and MIG — metallic, inert gas.

The TIG type uses an arc which is formed by a tungsten electrode fitted in the torch. Unlike the normal metal arc process, this electrode does not burn away to supply the weld with metal; instead a separate rod is used in conjunction with the torch in a manner similar to that used in oxy-acetylene welding.

The argon gas, stored under pressure in a cylinder painted blue, is supplied by a hose to the torch. When the operator opens the gas control valve on the torch, the gas is discharged around the electrode. The gas outlet nozzle (or ceramic shield) is cooled by water or air.

O_2 process

This process normally used equipment of the MIG type — the inert gas being carbon dioxide, which is supplied in a cylinder painted black.

The arc is produced by a bare metal electrode wire, fed from a reel to the torch at a rate governed by the welding current. The gas CO_2 is discharged around the electrode wire during the welding process and the flow is controlled by a valve on the torch.

Filler metal is provided by the electrode wire, which melts when it is dipped into the molten pool. Contact between metal and rod momentarily extinguishes the arc, and the increase in the welding current, due to the lower resistance, causes the tip of the electrode to melt off. These events are repeated many times a second.

.37 Resistance welding

The most common method of resistance welding used in M.V. repair establishments is 'spot welding' (Figure 2.37.1). This method relies on the heat generated when a large electrical current is passed through a metal. Since the current is localized to the spot where the two electrodes clamp the metal, the risk of serious distortion is overcome. Also, when this is combined with other advantages such as speed and ease of operation, then it will be seen why this method is adopted for fabrication of sheet metal panels used in vehicle body parts (Figure 2.37.2).

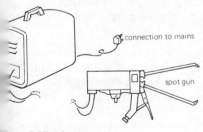

connection to mains

spot gun

ure 2.37.1 Spot welder

copper electrode

pushed together

Figure 2.37.2 Spot welding

elding heat production

The heat produced at the weld is governed by the
1. electrical resistance of the metal being welded
2. clamping pressure applied by the electrodes to the metal
3. current flow through the metal
4. period of time that the current flows

ectrical resistance of metals

A metal such as mild steel has a higher resistance than aluminium, so for a given current, mild steel will heat up quicker.

Clamping pressure

Good electrical contact must be made between the electrode and the metal, and between the metal sheets being welded. To achieve this contact, pressure must be applied.

Current flow

A welding transformer, normally connected to the mains, lowers the voltage and raises the current to the high value required, e.g. 10 000 at 1 V is used when welding two pieces of mild steel of thickness 1 A switch enables the operator to set the output current to suit the j

Time

Sufficient time must be allowed for the current to heat the metal to the temperature required for fusion, but as soon as this temperature is reached the current must be switched off to avoid the molten me being squeezed out by the electrodes. An automatic control which cuts off the current when it senses the correct temperature is ideal, failing this the next best thing is a timing control, as this enables the operator to set the period for current flow. In practice, the setting this timing control may result in some spot welds being imperfect, s to allow for this, extra welds are made.

2.38 Hard soldering, brazing and bronze welding

Hard soldering

The process of hard soldering, or silver soldering, is used when the requirement demands that the joint shall be either stronger or have the ability to withstand higher temperatures than that associated w soft soldering.

A typical silver solder has a melting temperature of about 700 °C and has a composition of

silver 60 per cent, copper 30 per cent, zinc 10 per cent

The method of preparation is similar to that applied to soft soldering and the flux normally used is borax. A soldering-iron cannot be used for hard soldering, instead the heat is applied direct onto the joint by means of brazing torch, or by means of an oxy-acetylene welding torch set to an oxidizing flame.

Brazing

Brazing is a metal joining process, performed at a temperature of about 900 °C, that uses a spelter or joining alloy consisting of copp and zinc. Various spelters are offered to suit different jobs — a typ spelter suitable for brazing mild steel has a composition of 65 per copper and 35 per cent zinc.

A joint to be brazed should have a small clearance between the mating parts to allow for capillary flow of the spelter. After cleani to the standard required for soldering, the surfaces are coated with borax paste (formed by mixing borax with water) and heated by a torch just sufficient to melt the spelter. Before this temperature is reached, the flux will have produced a film on the metal surface to prevent the air forming an oxide. If an oxide had formed, it would act as a barrier between the spelter and metal surface and result in poor joint. When the metal cools, the flux sets and produces a har deposit. This deposit can be removed by pickling (immersing the workpiece) in a weak solution of sulphuric acid or by sprinkling sa on the surface of the metal whilst it is still hot.

Bronze welding is performed with oxy-acetylene welding equipment and a bronze filler rod. With this process it is possible to make strong joints in copper and steel, join dissimilar metals and make repairs to cast iron components. Unlike normal fusion welding, bronze welding does not involve the melting of the parent metal; instead it joins the two surfaces with bronze. Bronze melts at a comparatively low temperature (900 $^\circ$C) so distortion is minimized. This advantage, together with 'high strength', makes the method attractive.

Various bronze welding rods are offered to suit different materials, e.g.

1. silicon bronze for mild steel
2. nickel bronze for mild steel where extra strength is required
3. manganese bronze for cast iron

A flux is required for bronze welding and a borax type is used. For convenience some filler rods are coated with flux.

Preparation involves the rounding off of any sharp edges and the cleaning of the metal surfaces in the vicinity of the join. A blowpipe nozzle about two sizes smaller than that used for normal welding is set to give an oxidizing flame.

In contrast to brazing, bronze welding requires the heat to be applied quickly and locally, so as applied to a lap joint, no large capillary flow is expected. The direction and angles used for leftward welding are adopted, but care must be taken to avoid melting the parent metal. When the metal is raised to a red heat, the bronze will run ahead of the main pool to coat and prepare the surface for the main deposit. Before and during the 'weld' the flux must be applied to ensure the surface is free from oxide contamination; periodic dipping of the hot rod into the powdered flux is the method used.

Great strength can be obtained from a bronze weld when the joint is in shear, so if the edges of the metal are unbevelled, then the deposit should be built up to overlap the joint.

Fractured castings should be preheated to about 450 $^\circ$C and after welding should be allowed to cool as slow as possible.

.39 Adhesives

Up to about fifty years ago the main adhesives were animal glues. About that time, vegetable glues were introduced for binding porous materials such as paper. The major development in adhesives came in the 1930s with the introduction of synthetic resins. These early resin-based adhesives had an excellent resistance to moisture and to mould growth and were particularly suitable for bonding wood. When aircraft constructions changed from wood to metal, adhesives were developed for metal bonding and from these many others were developed to suit other materials. From the wide range of modern adhesives it is normally possible to find one which not only suits the materials to be joined but also meets the operational requirements.

1. Appearance—the joint hardly shows
2. Strength—shearing force is spread over a large area instead of acting in local spots as in the case of a rivet
3. Reduced distortion—excessive heating is unnecessary
4. Corrosion—electrolytic action between dissimilar materials is reduced

Also there are advantages associated with particular application e.g. compared with rivets a bonded brake pad or lining has a larger friction area; this results in a lower rate of wear and improved fade characteristics.

Disadvantages

1. Health hazard—dangerous fumes are given off by many non-inflammable adhesives, so good ventilation is essential.
2. Fire and explosive risks—some adhesives give off an inflamm able vapour which becomes explosive when used in a confined spa
3. Temperature limitations—the bond will break when the tem ature exceeds the recommended figure.
4. Inspection difficulties—it is difficult to ascertain by visual inspection the strength of the bond.
5. Cost—expensive equipment is necessary for some special bor applications; consequently, the method is sometimes uneconomica

Technical terms applied to adhesives

The following terms are used in connection with adhesives:
Thermo-plastic: can be repeatedly softened by heat.
Thermo-setting: sets by the action of heat or a catalyst to a permanently hard state.
Impact type: an adhesive which is applied to both surfaces and allowed to become tacky. Contact between the surfaces produces the bond.
Cold set: an adhesive which is cured or set at room temperatur
Hot set: an adhesive which requires heating to a given tempera to complete the bond.
Structural adhesives: those suitable for applications where the bonded joint sustains continuous loads in service. Thermo-plastic adhesives are normally unsuitable for these applications since the soon fail under a continuous load.

Types of adhesive
Structural adhesives

Synthetic resins (thermosetting).
(a) Epoxy —normally a syrupy liquid, which, when mixed with a hardener (a catalyst, curing agent) is rapidly transformed into a h transparent solid. It bonds not only absorbent materials such as v but also adheres to metal and glass. Used for a wide range of M.V work which involves the bonding of metals, ceramics, glass, rubb plastics, wood, etc. Also used for glass fibre work.
(b) Phenolic—when mixed with epoxy or nitrile synthetic rubber these hot set resins have a high shear strength and withstand temp atures up to about 250 °C. Used for heavy duty applications suct bonding the friction material to steel brake shoes, pads or autom transmission clutches.

Non-structural adhesives

1. Elastomeric (rubber based).
(a) Natural rubber — for bonding rubber to rubber (e.g. tyre rep and fabric and leather.
(b) Natural rubber latex — for fabrics, leather, felt, paper etc.
(c) Synthetic rubber — for PVC, wood and glass
(d) Rubber and resin — for rubber, felt, cork, etc., to wood or m
(e) Synthetic rubber and resin—for laminated plastics, plywood, hardboard to metal or wood. Particularly suitable for smooth sur

2. Shellac. A natural resin used as an engine jointing compound for metal to metal surfaces or fibrous gasket material to metal surfaces. It is resistant to hydrocarbons at high temperatures.

3. Anaerobic (absence of air). Acrylic acid — one product sets when in contact with metal if oxygen is absent. Used as a 'liquid lock-washer' for bolts, it is capable of withstanding temperatures up to 200 °C; above 250 °C the material softens, so this feature is useful when breaking a joint.

.40 Multiple choice uestions

1. Which one of the following is the carbon composition of cast iron?
    (a)   0·15 per cent
    (b)   0·5 per cent
    (c)   1·5 per cent
    (d)   4 per cent

2. Which one of the following is the carbon composition of medium carbon steel?
    (a)   0·05—0·25 per cent
    (b)   0·25—0·55 per cent
    (c)   0·55—0·9 per cent
    (d)   0·9—4 per cent

3. A high carbon steel is heat treated to improve its ability to withstand scratching and identation. This process is performed by heating the steel to a 'cherry red' and
    (a)   cooling as slow as possible
    (b)   quenching in oil or water
    (c)   placing it into hot sand
    (d)   plunging it into a special compound

4. Steel at a temperature of 800 °C has a colour described as
    (a)   white hot
    (b)   cherry red
    (c)   dull red
    (d)   black heat

5. The cooling operation in the hardening process is called
    (a)   quenching
    (b)   dipping
    (c)   plunging
    (d)   tempering

6. Oil or water is used to absorb the heat during the hardening process. One advantage of oil is that the
    (a)   higher boiling point allows the cooling rate to be increased
    (b)   vaporization of the liquid increases the heat capacity
    (c)   rate of cooling is quicker and distortion is reduced
    (d)   rate of cooling is slower and this reduces the risk of cracking

7. A steel is heated to a temperature of 200 °C and suddenly cooled. This heat treatment process is called
    (a)   hardening
    (b)   normalizing
    (c)   tempering
    (d)   annealing

8. The effect of annealing a material is to
    (a)   reduce its hardness
    (b)   reduce its softness
    (c)   increase its resistance to wear
    (d)   increase its resistance to indentation

9. Working steel in a red-hot
   state for a long period of
   time causes the grain to
   become large and coarse.
   To restore the grain structure
   the steel should be
   (a) re-hardened
   (b) tempered
   (c) normalized
   (d) nitrided

10. The temperature of a material
    can be estimated from the
    colour of the oxide film
    which appears on a polished
    surface. This method is often
    used when
    (a) hardening
    (b) tempering
    (c) normalizing
    (d) annealing

11. Which one of the following
    oxide colours indicates the
    temperature at which quench-
    ing is necessary for a cutting
    tool subjected to severe
    impact or shock?
    (a) Pale straw
    (b) Dark straw
    (c) Dark purple
    (d) Blue

12. After hardening a steel
    cutting tool the material is
    too brittle to use, so to
    improve its shock resistance
    the tool is
    (a) annealed
    (b) tempered
    (c) normalized
    (d) case hardened

13. The process in which a steel
    is heated to a 'cherry red' and
    then allowed to cool as slow
    as possible is called
    (a) hardening
    (b) case hardening
    (c) tempering
    (d) annealing

14. If a component, such as a
    gudgeon pin, needs a toug
    core and a wear-resistant
    surface, then the heat trea
    ment required is called
    (a) hardening
    (b) tempering
    (c) annealing
    (d) case hardening

15. Heating steel parts in a bo
    packed with a carbon-rich
    substance is a method of
    (a) nitriding
    (b) normalizing
    (c) annealing
    (d) case hardening

16. The process in which spec
    alloy steel components m
    be surface hardened by he
    in contact with ammonia
    gas is called
    (a) nitriding
    (b) box
    (c) cyanide
    (d) open-hearth

17. Which one of the followir
    has the largest percentage
    of 'free carbon'?
    (a) Mild steel
    (b) Cast iron
    (c) Alloy steel
    (d) High carbon steel

18. When steel is 'cold worke
    the crystal structure is
    distorted and this makes
    the steel
    (a) softer and impact
         resistant
    (b) softer and increases
         the strength
    (c) harder and more br
    (d) harder and impact
         resistant

19. Which one of the followi
    materials is suitable for c
    hardening?
    (a) Low carbon steel
    (b) High carbon steel

(c) Aluminium
(d) Copper

(a) Brass
(b) Nickel
(c) Bronze
(d) Y metal

20. The effect of overheating a
clutch spring is that it will
(a) make it harder
(b) make it more brittle
(c) increase the thrust that
it applies on the plate
(d) decrease the thrust
that it applies on the
plate

21. Which one of the following
is a non-ferrous alloy?
(a) Brass
(b) Copper
(c) Nickel-chrome steel
(d) Chrome vanadium steel

22. A material which becomes
much harder when it is
heated to a 'cherry-red' and
quenched in water is
(a) low-carbon steel
(b) high-carbon steel
(c) copper
(d) aluminium

23. A strip of copper is bent
backwards and forwards a
number of times. This
action makes the copper
(a) harder
(b) softer
(c) more ductile
(d) more malleable

24. A strip of copper is heated
to a dull red heat and
quenched in water. This
process causes the copper
to become
(a) harder
(b) more brittle
(c) more ductile
(d) less ductile

25. Which one of the following
is a common alloying
element that is used to
improve the toughness of
a steel?

26. Which one of the following
is an alloying element that
improves the resistance of a
steel to corrosion?
(a) Vanadium
(b) Tungsten
(c) Manganese
(d) Chromium

27. What is Mazac?
(a) A ferrous alloy
(b) A zinc based alloy
(c) A solder consisting of
60 per cent tin and
40 per cent lead
(d) A common crank-
shaft bearing material

28. Brass is an alloy consisting
of copper and
(a) tin
(b) lead
(c) zinc
(d) aluminium

29. Bronze is an alloy consisting
of copper and
(a) tin
(b) lead
(c) zinc
(d) aluminium

30. Which one of the following
is a light alloy which 'age-
hardens'?
(a) Tin
(b) Zinc
(c) Aluminium
(d) Duralumin

31. A load of 400 N acts on a
rod having a cross-sectional
area of 200 mm^2. The stress
is
(a) 0·5 N/mm^2
(b) 1 N/mm^2
(c) 2 N/mm^2
(d) 80 kN/mm^2

32. A load of 400 N acts on a rod having a cross-sectional area of 200 mm^2. The stress is
    (a) 0·5 MN/m^2 or 0·5 MPa
    (b) 1 MN/m^2 or 1 MPa
    (c) 2 MN/m^2 or 2 MPa
    (d) 80 GN/m^2 or 80 GPa

33. During the power stroke the force in the connecting rod is 6 kN. What is the magnitude and type of stress at the mid point in the connecting rod where the cross-section has an area of 120 mm^2?
    (a) 50 MPa ; compressive
    (b) 20 GPa ; compressive
    (c) 50 MPa ; tensile
    (d) 20 GPa; tensile

34. Corrosion has reduced the diameter of a circular brake rod by 50 per cent. During brake application the stress in the rod will be
    (a) half the original stress
    (b) twice the original stress
    (c) four times the original stress
    (d) eight times the original stress

35. The cross section area of a gudgeon pin is 120 mm^2. When the load on the pin is 6 kN the stress in the pin is
    (a) 25 N/mm^2
    (b) 50 N/mm^2
    (c) 20 kN/mm^2
    (d) 40 kN/mm^2

36. 'Provided the load on a steel bolt does not exceed the _____, the material will return to its original length when the load is removed.' The words needed to complete this sentence are
    (a) yield point
    (b) breaking load
    (c) factor of safety
    (d) ultimate tensile lo

37. The colour of an oxyge cylinder and the directi of the threads used to connect the oxygen equipment is
    (a) maroon and left-h
    (b) maroon and right
    (c) black and left-har
    (d) black and right-ha

38. Leakage of gas from ox equipment can be trace using
    (a) oil
    (b) grease
    (c) powdered graphit
    (d) soapy water

39. Which one of the follov should NOT be used on gas welding equipment
    (a) Oil or grease
    (b) Powdered graphi
    (c) Soapy water
    (d) Acetone

40. The colour of an acety cylinder and the directi of the threads used to connect combustible ga equipment is
    (a) maroon and left-
    (b) maroon and righ
    (c) black and left-ha
    (d) black and right-h

41. Why should the acetyle cylinder only be used v it is positioned vertical with the valve at the to
    (a) To prevent the c overheating
    (b) To enable the va be shut off quick
    (c) To avoid liquid a entering the pipe
    (d) To avoid kapoc the regulator

42. The purpose of the regulator in oxy-gas welding equipment is to
    (a) lower the pressure
    (b) prevent a flash back
    (c) restrict the quantity of gas
    (d) control the temperature of the welding flame

43. The size of the orifice fitted to a welding torch is indicated by
    (a) a colour code
    (b) a number stamped on its side
    (c) the diameter of the shank
    (d) the length of the nozzle stem

44. A low setting of the gas regulators can cause
    (a) backfiring
    (b) a low flame temperature
    (c) overheating of the blowpipe
    (d) the gas velocity at the nozzle to be excessive

45. If a flame is given off from a hose or valve of the welding equipment, then
    (a) evacuate the premises immediately
    (b) saturate the complete system with water
    (c) open both cylinder valves fully to extinguish the flame
    (d) the acetylene cylinder valve should be turned off immediately

46. A welding flame having a 'whitish' feather which appears beyond the inner luminous cone is called
    (a) carbonizing
    (b) natural
    (c) neutral
    (d) oxidizing

47. An oxy-acetylene welding process which used a neutral flame and a leftward technique is recommended for
    (a) brazing
    (b) bronze welding
    (c) thin steel plate
    (d) thick steel plate

48. The recommended blowpipe and rod angles for butt welding two plates of mild steel of thickness 4 mm are
    (a) blowpipe $30°$, rod $60°$
    (b) blowpipe $40°$, rod $50°$
    (c) blowpipe $50°$, rod $40°$
    (d) blowpipe $60°$, rod $30°$

49. Which one of the following requires a flux to prevent the metal becoming oxidized when being gas welded?
    (a) Thin mild steel plate
    (b) Fillet welding of thin steel plate
    (c) Butt welding of thick steel plate
    (d) Bronze welding of steel plate

50. Undercutting during a fillet weld can be avoided if
    (a) the gas regulator is adjusted to give a smaller flame
    (b) the blowpipe is directed into the 'V' to form a trap for the heat
    (c) a smaller flame than that required for butt welding is directed towards the smallest mass of metal
    (d) a larger flame than that required for butt welding is directed towards the greatest mass of metal

51. Distortion during gas-welding thin steel plates is minimized by
    (a) using a large flame
    (b) tacking the plates at regular intervals
    (c) preheating the complete plate
    (d) cooling the under-side of the weld

52. Oxygen used in the oxy-gas cutting process has two duties; it blows out the molten metal and also
    (a) oxidizes the iron to lower the melting point
    (b) melts the steel by heating it beyond the welding temperature
    (c) causes the metal to form a combustible substance
    (d) gives a local heat to allow the acetylene to react with the iron

53. What voltage is required from a metal arc welding transformer to 'strike the arc'?
    (a) 10–25 V
    (b) 60–90 V
    (c) 130–170 V
    (d) 210–250 V

54. What is meant by 'arc-blow' as applied to metal arc welding equipment?
    (a) The failure of the current to maintain an arc
    (b) The scatter of red-hot particles on each side of the weld
    (c) A stream of inert gas supplied from a torch to reduce oxidation
    (d) When the arc wanders due to the presence of a magnetic field

55. Polarity of an electrode used with a d.c. welding generator affects the rate in which the electrode rod is 'burnt', because the
    (a) current must flow from the electrode to earth to strike the arc
    (b) current must flow from earth to the electrode to maintain the arc
    (c) heat generated at the rod depends on the polarity
    (d) heat generated at the positive pole is always lower

56. One purpose of the thick coating on an electrode rod is to
    (a) prevent rusting
    (b) increase the rod diameter and reduce the electrical resistance
    (c) prevent the air from contacting the hot metal
    (d) remove the oxide from the surface before welding

57. Goggles used for oxy-acetylene welding are unsuitable for metal arc-welding because the goggles will not
    (a) resist the intense heat
    (b) direct the fumes away from the face
    (c) prevent the spatter from entering the eye
    (d) protect the eyes and face from the dangerous rays

58. If the electrode is moved too fast when performing a metal arc-weld, the result is that the
    (a) penetration will be poor
    (b) crater will be too deep

(c) electrode becomes
red hot

(d) deposit of metal will
be excessive

59. If the electrode is moved too
slow when performing a
metal arc-weld, the result is
that the
(a) penetration will be
poor
(b) crater will be too
deep
(c) parent metal will not
be melted
(d) deposit of metal will
be insufficient

60. A metal arc-weld has to be
stopped to change a rod
before the complete 'run'
has been finished. Before
striking the arc what operation is necessary?
(a) The polarity of the
rod should be changed
(b) The flux must be
removed from the
finished weld before
the metal cools
(c) The crater must be
filled with chippings
obtained from the
rod coating
(d) The slag is chipped
from the end of the
bead and the crater
is cleaned

61. Why are special precautions
necessary when electric
welding is performed on a
vehicle fitted with semi-
conductor devices? The
devices
(a) are ruined when sub-
jected to heat or
induced currents
(b) generate an electric
current which can be
harmful to human
beings

(c) explode if they are
exposed to a magnetic
field

(d) produce a magnetic
field which causes
'arc-blow'

62. As applied to welding, the
term 'shielded-arc' means
that the arc is
(a) contained in a flux
melted from the rod
covering
(b) covered by a screen
incorporated in the
electrode holder
(c) enclosed in an envelope
of inert gas
(d) screened by fireproof
curtains to avoid
injury to personnel

63. As applied to welding, the
term 'MIG' is an abbrevi-
ation for
(a) medium industrial gas
(b) metallic inert gas
(c) maximum ideal gauge
(d) minimum insulation
group

64. A gas cylinder painted blue
is used in conjunction with
a welding transformer. What
is the type of gas and what
is its purpose?
(a) Argon gas which is
used to prevent the
air from affecting
the weld
(b) Argon gas which is used
to cool the welding
torch
(c) Carbon dioxide which
is used to resist the
formation of nitrides
(d) Carbon dioxide which
is used to prevent the
action of the oxides

65. One type of resistance welding is called
    (a) spot
    (b) inert
    (c) friction
    (d) metal arc

66. Two factors affecting the heat produced by resistance welding are
    (a) penetration and current flow
    (b) current flow and depth of crater
    (c) electrical resistance of the material and depth of crater
    (d) clamping pressure and period of time that current flows

67. Resistance welding is often used to join motor vehicle body parts because
    (a) distortion is minimized
    (b) heat can be applied to a large area
    (c) each weld is fused correctly
    (d) corroded panels can be welded satisfactorily

68. Which one of the following processes is suitable for joining two pieces of steel with the minimum of heating if the steel has to withstand a temperature of 700 °C when in service?
    (a) Silver soldering
    (b) Soft soldering
    (c) Bronze welding
    (d) Fusion welding

69. A suitable flux for brazing mild steel is
    (a) resin
    (b) killed spirits
    (c) borax
    (d) weak solution of sulphuric acid or salt

70. What type of oxy-acetylene flame and technique is recommended for bronze welding of mild steel?
    (a) Oxidizing and leftward
    (b) Oxidizing and rightward
    (c) Carbonizing and leftward
    (d) Carbonizing and rightward

71. What size of nozzle, compared to normal fusion welding, and what type of welding rod is recommended for bronze-welding an exhaust manifold of an engine?
    (a) Nozzle about two sizes smaller and silicon bronze rod
    (b) Nozzle about two sizes smaller and manganese bronze rod
    (c) Nozzle about two sizes larger and silicon bronze rod
    (d) Nozzle about two sizes larger and manganese bronze rod

72. Which one of the following is a thermosetting synthetic adhesive?
    (a) Latex
    (b) Shellac
    (c) Epoxy
    (d) Anaerobic

.50 Ohm's law

The current passing through a wire at constant temperature is proportional to the potential difference between its ends

This law was discovered by Dr G. S. Ohm in 1826 and is now known as Ohm's Law.

From the law the following expression can be formed.

A potential difference of 1 volt is required to produce a current of 1 ampere through a conductor of resistance 1 ohm. Writing this in symbol form

$$V = IR$$

where V = p.d. (volt)

I = current (ampere)

R = resistance (ohm)

therefore

$$I = \frac{V}{R}$$

and

$$R = \frac{V}{I}$$

Example 1

What voltage is required to produce a current of 6 A through a resistance of 1·5 Ω?

$$V = IR$$
$$V = 6 \times 1\cdot5$$
$$V = 9 \text{ V}$$

Example 2

A current of 3 A flows through the primary winding of an ignition coil when it is connected to a battery having a p.d. of 12 V. Find the resistance of the coil winding.

$$V = IR$$

$$R = \frac{V}{I}$$

$$= \frac{12}{3}$$

$$R = 4 \ \Omega$$

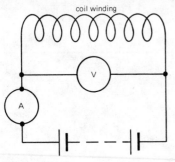

coil winding

Figure 2.50.1

This calculation shows a simple method of testing the windings of any component — a short circuit would give a lower resistance value, whereas a bad connection within the unit would give a high resistance. Comparing the result with the value given by the manufacturer enables the condition of the windings to be determined. Figure 2.50.1 shows the arrangement of the circuit for measuring the resistance of a coil winding.

Example 3

Determine the current flowing in a circuit of resistance 2 Ω when the supply p.d. is 12 V.

$$V = IR$$

$$I = \frac{V}{R}$$

$$= \frac{12}{2}$$

$$I = 6 \text{ A}$$

2.51 Voltage distribution in circuits

Voltage distribution — resistors in series

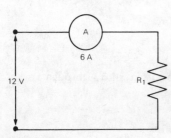

Figure 2.51.1

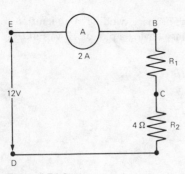

Figure 2.51.2

Having seen the relationship between p.d., current flow and resistan we can now investigate the distribution of voltage in a circuit.

To simplify the following examples, it is assumed that the cable offers no resistance to the flow of current.

Consider a circuit (Figure 2.51.1) consisting of a resistor and amme which is connected to a 12 V supply. If the reading on the ammete 6 A then the value of the resistor R_1 would be

$$R_1 = \frac{V}{I}$$

$$R_1 = \frac{12}{6} = 2 \text{ Ω}$$

When a 4 Ω resistor R_2 is connected in series with the first resis (Figure 2.51.2) it is found that the current flow is now 2 A. The t resistance of this circuit is:

$$R = \frac{V}{I}$$

$$R = \frac{12}{2} = 6 \text{ Ω}$$

OR

$$R = R_1 + R_2$$

This shows that

The total resistance of a circuit comprising a number of resistors connected in series is obtained by adding the values of all resistors

The p.d. at different parts of the circuit in Figure 2.51.2 may be determined by using a voltmeter, e.g. a meter connected to points B and C shows a p.d. of 4 V. This reading shows that the voltage has decreased by 4 V after the current has passed through the resistor R_1. By connecting the meter to points C and D the reading should be 8 V. These results may be summarized in tabular form.

Position of meter leads	Reading (V)
E.D.	12
B.C.	4
C.D.	8

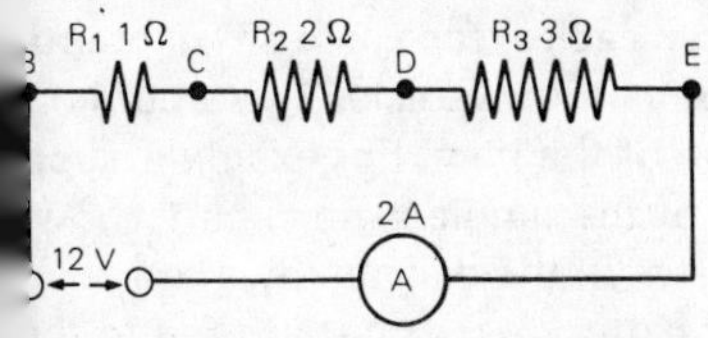

Figure 2.51.3 shows three resistors having values of 1 Ω, 2 Ω and 3 Ω respectively, connected in series to a 12 V supply. If the current taken from the supply is 2 A, then

$$\text{Total resistance of circuit} = R_1 + R_2 + R_3$$
$$= 1 + 2 + 3$$
$$= 6\ \Omega$$

Figure 2.51.3

Connecting a voltmeter between B and C would give a p.d. of 2 V; this p.d. is often called the *voltage-drop* because the resistor has the effect of lowering the potential. This drop can also be calculated from

$$V = IR$$

therefore

p.d. across resistor = current x resistance
p.d. across 1 Ω resistor = 2 x 1 = 2 V
p.d. across 2 Ω resistor = 2 x 2 = 4 V
p.d. across 3 Ω resistor = 2 x 3 = 6 V
Total p.d. = 12 V

The total must equal the supply p.d.

It will be noted that the same current flows through each resistor, i.e. the current flowing through any part of a series circuit is the same.

The example shows the relationship between the total voltage applied across a number of series-connected circuit elements and the voltage across the individual elements. The ratio between the p.d. across each resistor and the supply p.d. is the same as the ratio between the individual resistance value and the total resistance of the external circuit. In other words

$$\frac{\text{p.d. across resistor}}{\text{p.d. of supply}} = \frac{\text{individual resistance value}}{\text{total resistance of external circuit}}$$

Table 12 indicates voltmeter readings obtainable from the circuit shown in Figure 2.51.3.

Table 12. Voltage across circuit elements in Figure 2.51.3

Position of meter leads	Reading (V)
B.C.	2
C.D.	4
D.E.	6
B.D.	6
C.E.	10

Practical applications

Knowledge of the behaviour of resistors mounted in series enables the student to understand various practical applications.

Example 1

There are occasions when a lamp of low voltage, e.g. an ignition warning lamp, is fitted to a higher voltage system. Subjecting the lamp to full battery p.d. would quickly destroy the bulb, so a resistor is connected in series with the lamp to drop the p.d. to the value required by the lamp.

Example 2

Terminal corrosion, or partly broken cables, can cause an unintenti resistance in the electrical circuit of a motor vehicle. This would dr the voltage and affect the operation of the item. For example, supp a resistance develops in the switch of the simple lamp circuit, show in Figure 2.51.4. Replacing the switch in the diagram by a resistor (Figure 2.51.5) shows that the full battery p.d. is not applied to the lamp; the drop in p.d. depends on the value of the resistor.

To locate the fault quickly, a voltmeter could be used. With the switch closed, the voltmeter should show battery p.d. at points 1 t 3, but in this case the reading at 3 will be considerably lower.

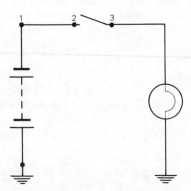

Figure 2.51.4

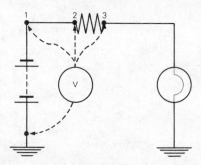

Figure 2.51.5

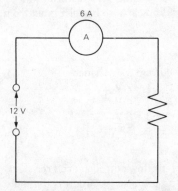

Figure 2.51.6

Voltage distribution — resistors in parallel

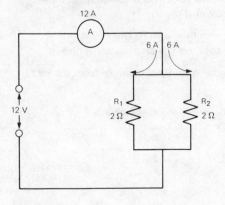

Figure 2.51.7

In Figure 2.51.6 a single resistor is connected to a 12 V supply. To give a current flow of 6 A, the circuit would require a resistance value of

$$R = \frac{V}{I}$$

$$= \frac{12}{6}$$

$$= 2\ \Omega$$

If another 2 Ω resistor is fitted in parallel with R_1 (Figure 2.51 then the current that would flow through the ammeter would incr to 12 A. Study of this circuit shows that the additional resistor of the current an alternative path, so a larger current would flow fro the supply. A 2 Ω resistor will pass a current of 6 A when the p.d. 12 V, and since this p.d. applies to both resistors, then the total c must be 6 + 6 = 12 A.

The resistance of the circuit will be

$$R = \frac{V}{I}$$

$$= \frac{12}{12}$$

$$= 1 \ \Omega$$

Therefore, two resistors of 2 Ω each give an equivalent resistance of 1 Ω. In other words, the two resistors are equivalent to a resistor having a value of 1 Ω.

Figure 2.51.8 shows two resistors of unequal value connected in parallel and an ammeter showing a current flow of 6 A.

Both resistors are subject to a p.d. of 12 V, and in order to find the current flowing through each, use the expression

$$I = \frac{V}{R}$$

Current flowing through $R_1 = \frac{12}{3} = 4 \ A$

Current flowing through $R_2 = \frac{12}{6} = 2 \ A$

Total current $= 6 \ A$

Obviously the greater current flows through the resistor of lower value. Thus the 3 Ω resistor carries 4/6 of the total current, and the 6 Ω resistor carried 2/6 of the total current.

Equivalent resistance of the circuit is

$$R = \frac{V}{I} = \frac{12}{6} = 2 \ \Omega$$

Let us now consider the result of connecting three resistors in parallel. Figure 2.51.9 shows a circuit consisting of three branches incorporating resistors of values 2 Ω, 3 Ω and 6 Ω. When the circuit is connected to a 12 V supply the current flow through

$$R_1 \text{ is } \frac{V}{R_1} = \frac{12}{2} = 6 \ A$$

$$R_2 \text{ is } \frac{V}{R_2} = \frac{12}{3} = 4 \ A$$

$$R_3 \text{ is } \frac{V}{R_3} = \frac{12}{6} = 2 \ A$$

Total current flowing in circuit = 12 A

Equivalent resistance of circuit $= \frac{V}{I} = \frac{12}{12} = 1 \ \Omega$

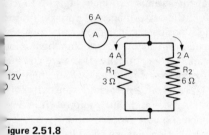

Figure 2.51.8

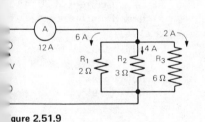

Figure 2.51.9

Fall of potential

All materials resist the flow of electrical current. Even a good copper cable causes a fall in potential as the current passes along the cable. This is illustrated by considering a starter cable having a resistance of 0·0006 Ω per metre. When a typical starter current of 200 A is supplied by the cable, the drop in potential (volt drop) per metre length of cable is

$$V = IR$$
$$= 200 \times 0.0006$$
$$= 0.12 \text{ V}$$

This does not seem much, but if the battery is positioned at a point which required 5 metres of cable, then the drop in potential amount to 0.6 V. A cable drop of this extent added to the numerou other 'volt-drops' occurring at the various terminals, adds up to a value which can seriously affect the performance of a starter motor

In this example, cable length has been considered, but cross-sectional area must also be taken into account; reducing the numbe of strands increases the resistance proportionally. If the area of the starter cable in the previous example was reduced by half, then the voltage drop would then be 1.2 V. Heat generated by the current would also cause the cable to overheat and this would result in a further increase in the resistance and volt drop.

Further examples of electrical principles are given in *F. of M.V.*

2.52 Multiple choice questions

1. Ohm's law states: 'The current passing through a wire at constant temperature is proportional to the
   (a) power supplied'
   (b) length of the circuit'
   (c) resistance of the circuit'
   (d) potential difference between its ends'

2. The potential difference that is required to cause a current of 8 A to flow through a resistance of 2 Ω is
   (a)   0.25 V
   (b)   2 V
   (c)   4 V
   (d)   16 V

3. A current of 6 A flows through the primary winding of an ignition coil when it is connected to a battery having a p.d. of 12 V. The resistance of the winding is
   (a)   0.5 Ω
   (b)   2 Ω
   (c)   6 Ω
   (d)   72 Ω

4. A voltage of 20 V is applie to a circuit of resistance 2 The current flow is
   (a)   0.1 A
   (b)   1 A
   (c)   10 A
   (d)   40 A

5. What voltage is required to cause a current of 20 mA t flow through a resistance c 40 Ω?
   (a)   0.5 mV
   (b)   0.8 V
   (c)   2 kV
   (d)   80 kV

6. Two field windings of equ resistance are connected i series with a battery of p.c 12 V. If the current is 2 A the resistance of EACH winding is
   (a)   3 Ω
   (b)   6 Ω
   (c)   12 Ω
   (d)   24 Ω

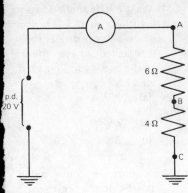

Figure 2.52.1

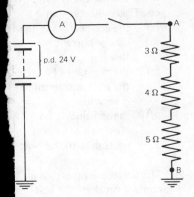

Figure 2.52.2

Questions 7–12 apply to Figure 2.52.1

7. The diagram shows an
    (a) earth return circuit consisting of two resistors in series
    (b) earth return circuit consisting of two resistors in parallel
    (c) insulated return circuit consisting of two resistors in series
    (d) insulated return circuit consisting of two resistors in parallel

8. Assuming the resistance of the connecting wires is zero, the total resistance of the circuit is
    (a) 0·7 Ω
    (b) 1·5 Ω
    (c) 2·4 Ω
    (d) 10 Ω

9. The current flow in the circuit is
    (a) 2 A
    (b) 8 A
    (c) 13 A
    (d) 15 A

10. The current flowing through the 6 Ω resistor is
    (a) 1·7 A
    (b) 2 A
    (c) 4 A
    (d) 10 A

11. A voltmeter connected to points A and B will register
    (a) 4 V
    (b) 6 V
    (c) 12 V
    (d) 20 V

12. The voltage-drop across the 4 Ω resistor is
    (a) 4 V
    (b) 6 V
    (c) 8 V
    (d) 20 V

Questions 13–17 apply to Figure 2.52.2

13. The diagram shows a circuit with three resistors in
    (a) series and the battery negative earthed
    (b) series and the battery positive earthed
    (c) parallel and the battery negative earthed
    (d) parallel and the battery positive earthed

14. The resistors have a total resistance of
    (a) 1·3 Ω
    (b) 2 Ω
    (c) 8 Ω
    (d) 12 Ω

15. The current flow in the circuit is
    (a) 0·5 A
    (b) 2 A
    (c) 19 A
    (d) 24 A

16. The voltage-drop across the 4 Ω resistor is
    (a) 2 V
    (b) 3 V
    (c) 6 V
    (d) 8 V

17. If the switch contacts develop a resistance of 4 Ω what is the p.d. between A and B?
    (a) 1·5 V
    (b) 6 V
    (c) 16 V
    (d) 18 V

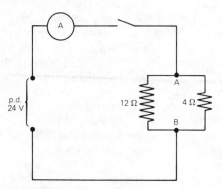

Figure 2.52.3

Questions 18–21 apply to Figure 2.52.3.

18. The resistance of the circuit is
    (a) 1·5 Ω
    (b) 3 Ω
    (c) 8 Ω
    (d) 16 Ω

19. The current flow in the circuit is
    (a) 1·5 A
    (b) 2 A
    (c) 6 A
    (d) 8 A

20. The voltage-drop across the 4 Ω resistor is
    (a) 6 V
    (b) 20 V
    (c) 24 V
    (d) 32 V

21. If the switch contacts develop a resistance of 9 Ω (i) what current will flow and (ii) what is the p.d. between A and B?
    (a) (i) 1 A (ii) 3 V
    (b) (i) 1 A (ii) 21 V
    (c) (i) 2 A (ii) 6 V
    (d) (i) 2 A (ii) 18 V

22. What resistor must be inserted in series with a lamp of resistance 6 Ω and battery of p.d. 12 V if the p.d. required at the lamp is 6 V?
    (a) 1 Ω
    (b) 2 Ω
    (c) 4 Ω
    (d) 6 Ω

23. What p.d. is applied to an ignition coil of 4 Ω resistance if a resistance of 2 Ω is connected between the coil and a battery having a p.d. of 12 V?
    (a) 2 V
    (b) 4 V
    (c) 6 V
    (d) 8 V

24. A voltmeter, connected in parallel to the positive battery terminal and positive post of a 12 V battery, shows a reading of 12 V when the lights are switched on. This indicates that
    (a) the connection is good
    (b) a large current is flowing
    (c) the resistance between the post and terminal is very low
    (d) the terminal is not making a good connection with the post

25. The voltage-drop across a contact-breaker of a coil-ignition system should not exceed 0·2 V. A voltage-drop greater than this limit causes the
    (a) current flow in the primary circuit to be reduced
    (b) current flow in the primary circuit to be increased
    (c) p.d. applied to the coil to be much larger
    (d) power applied to the coil to be much larger

3 TABLES AND ANSWERS

3.1 Conversion factors

Most of the following conversions have been approximated to give values suitable for general work. If greater accuracy is required, the internationally agreed equivalent should be used. To convert from Imperial to SI, the Imperial value is multiplied by the given conversion factor.

Length

1 yard is exactly 0·9144 metre

1 inch (in) ≏ 25·4 mm   1 foot (ft) ≏ 0·3048 m

0·001 in ≏ 0·025 mm   1 mile ≏ 1·6 km

Area

1 sq inch (in^2) ≏ 6·45 cm^2 1 sq foot (ft^2) ≏ 929 cm^2

Volume

1 cubic inch (in^3) ≏ 16·4 cm^3 1 cubic foot (ft^3) ≏ 0·028 m^3

Capacity

1 pint (pt) ≏ 0·568 litre   1 gallon (gal) ≏ 4·5 litre

Mass

1 pound (lb) is exactly 0·453 592 37 kg

1 ounce (oz) ≏ 28·35 g   1 ton ≏ 1016 kg = 1·016 tonne

Force

1 pound force (lbf) ≏ 4·45 N   1 ton ≏ 10 kN

Torque

1 pound foot (lbf ft) ≏ 1·4 N m

Pressure and stress

1 pound/sq inch (lbf/in^2) ≏ 7 kN/m^2 = 7 kPa = 0·07 bar = 70 mbar

1 ton/sq inch (ton/in^2) ≏ 15 kN/m^2 = 15 kPa

1 atmosphere (atm) ≏ 100 kN/m^2 = 100 kPa = 1 bar

1 inch mercury (in Hg) ≏ 3·4 kN/m^2 = 3·4 kPa = 34 mbar

Work and energy

1 foot pound (ft lbf) ≏ 1·4 J

1 British thermal unit (Btu) ≏ 1055 J = 1·055 kJ

1 Centigrade heat unit (Chu) ≏ 1900 J = 1·9 kJ

Power

1 horsepower (hp) ≏ 746 W = 0·746 kW

Velocity

1 foot/second (ft/s) ≏ 0·3 m/s

1 mile/h (mile/h) ≏ 1·6 km/h

Acceleration
1 foot/second^2 (ft/s^2) $\hat{=}$ 0.3 m/s^2
$g \hat{=} 32.2$ ft/s^2 $\hat{=}$ 9.81 m/s^2 $\hat{=}$ 10 m/s^2

Consumption
1 mile/gal $\hat{=}$ 0.35 km/litre or 290 litres/100 km
30 mile/gal $\hat{=}$ 11 km/litre or 9.7 litre/100 km
1 pt/bhp h $\hat{=}$ 0.7476 litres/kW h

3.2 Square root tables

	0	1	2	3	4	5	6	7	8	9	1	2	3	4	5	6	7	8	9
1.0	1.000	1.005	1.010	1.015	1.020	1.025	1.030	1.034	1.039	1.044	0	1	1	2	2	3	3	4	4
1.1	1.049	1.054	1.058	1.063	1.068	1.072	1.077	1.082	1.086	1.091	0	1	1	2	2	3	3	4	4
1.2	1.095	1.100	1.105	1.109	1.114	1.118	1.122	1.127	1.131	1.136	0	1	1	2	2	3	3	4	4
1.3	1.140	1.145	1.149	1.153	1.158	1.162	1.166	1.170	1.175	1.179	0	1	1	2	2	3	3	3	4
1.4	1.183	1.187	1.192	1.196	1.200	1.204	1.208	1.212	1.217	1.221	0	1	1	2	2	2	3	3	4
1.5	1.225	1.229	1.233	1.237	1.241	1.245	1.249	1.253	1.257	1.261	0	1	1	2	2	2	3	3	4
1.6	1.265	1.269	1.273	1.277	1.281	1.285	1.288	1.292	1.296	1.300	0	1	1	2	2	2	3	3	3
1.7	1.304	1.308	1.311	1.315	1.319	1.323	1.327	1.330	1.334	1.338	0	1	1	2	2	2	3	3	3
1.8	1.342	1.345	1.349	1.353	1.356	1.360	1.364	1.367	1.371	1.375	0	1	1	1	2	2	3	3	3
1.9	1.378	1.382	1.386	1.389	1.393	1.396	1.400	1.404	1.407	1.411	0	1	1	1	2	2	3	3	3
2.0	1.414	1.418	1.421	1.425	1.428	1.432	1.435	1.439	1.442	1.446	0	1	1	1	2	2	2	3	3
2.1	1.449	1.453	1.456	1.459	1.463	1.466	1.470	1.473	1.476	1.480	0	1	1	1	2	2	2	3	3
2.2	1.483	1.487	1.490	1.493	1.497	1.500	1.503	1.507	1.510	1.513	0	1	1	1	2	2	2	3	3
2.3	1.517	1.520	1.523	1.526	1.530	1.533	1.536	1.539	1.543	1.546	0	1	1	1	2	2	2	3	3
2.4	1.549	1.552	1.556	1.559	1.562	1.565	1.568	1.572	1.575	1.578	0	1	1	1	2	2	2	3	3
2.5	1.581	1.584	1.587	1.591	1.594	1.597	1.600	1.603	1.606	1.609	0	1	1	1	2	2	2	3	3
2.6	1.612	1.616	1.619	1.622	1.625	1.628	1.631	1.634	1.637	1.640	0	1	1	1	2	2	2	2	3
2.7	1.643	1.646	1.649	1.652	1.655	1.658	1.661	1.664	1.667	1.670	0	1	1	1	2	2	2	2	3
2.8	1.673	1.676	1.679	1.682	1.685	1.688	1.691	1.694	1.697	1.700	0	1	1	1	1	2	2	2	3
2.9	1.703	1.706	1.709	1.712	1.715	1.718	1.720	1.723	1.726	1.729	0	1	1	1	1	2	2	2	3
3.0	1.732	1.735	1.738	1.741	1.744	1.746	1.749	1.752	1.755	1.758	0	1	1	1	1	2	2	2	3
3.1	1.761	1.764	1.766	1.769	1.772	1.775	1.778	1.780	1.783	1.786	0	1	1	1	1	2	2	2	3
3.2	1.789	1.792	1.794	1.797	1.800	1.803	1.806	1.808	1.811	1.814	0	1	1	1	1	2	2	2	2
3.3	1.817	1.819	1.822	1.825	1.828	1.830	1.833	1.836	1.838	1.841	0	1	1	1	1	2	2	2	2
3.4	1.844	1.847	1.849	1.852	1.855	1.857	1.860	1.863	1.865	1.868	0	1	1	1	1	2	2	2	2
3.5	1.871	1.873	1.876	1.879	1.881	1.884	1.887	1.889	1.892	1.895	0	1	1	1	1	2	2	2	2
3.6	1.897	1.900	1.903	1.905	1.908	1.910	1.913	1.916	1.918	1.921	0	1	1	1	1	2	2	2	2
3.7	1.924	1.926	1.929	1.931	1.934	1.936	1.939	1.942	1.944	1.947	0	1	1	1	1	2	2	2	2
3.8	1.949	1.952	1.954	1.957	1.960	1.962	1.965	1.967	1.970	1.972	0	1	1	1	1	2	2	2	2
3.9	1.975	1.977	1.980	1.982	1.985	1.987	1.990	1.995	1.995	1.997	0	1	1	1	1	2	2	2	2
4.0	2.000	2.002	2.005	2.007	2.010	2.012	2.015	2.017	2.020	2.022	0	0	1	1	1	1	2	2	2
4.1	2.025	2.027	2.030	2.032	2.035	2.037	2.040	2.042	2.045	2.047	0	0	1	1	1	1	2	2	2
4.2	2.049	2.052	2.054	2.057	2.059	2.062	2.064	2.066	2.069	2.071	0	0	1	1	1	1	2	2	2
4.3	2.074	2.076	2.078	2.081	2.083	2.086	2.088	2.090	2.093	2.095	0	0	1	1	1	1	2	2	2
4.4	2.098	2.100	2.102	2.105	2.107	2.110	2.112	2.114	2.117	2.119	0	0	1	1	1	1	2	2	2
4.5	2.121	2.124	2.126	2.128	2.131	2.133	2.135	2.138	2.140	2.142	0	0	1	1	1	1	2	2	2
4.6	2.145	2.147	2.149	2.152	2.154	2.156	2.159	2.161	2.163	2.166	0	0	1	1	1	1	2	2	2
4.7	2.168	2.170	2.173	2.175	2.177	2.179	2.182	2.184	2.186	2.189	0	0	1	1	1	1	2	2	2
4.8	2.191	2.193	2.195	2.198	2.200	2.202	2.205	2.207	2.209	2.211	0	0	1	1	1	1	2	2	2
4.9	2.214	2.216	2.218	2.220	2.225	2.227	2.229	2.232	2.232	2.234	0	0	1	1	1	1	2	2	2
5.0	2.236	2.238	2.241	2.243	2.245	2.247	2.249	2.252	2.254	2.256	0	0	1	1	1	1	2	2	2
5.1	2.258	2.261	2.263	2.265	2.267	2.269	2.272	2.274	2.276	2.278	0	0	1	1	1	1	2	2	2
5.2	2.280	2.283	2.285	2.287	2.289	2.291	2.293	2.296	2.298	2.300	0	0	1	1	1	1	2	2	2
5.3	2.302	2.304	2.307	2.309	2.311	2.313	2.315	2.317	2.319	2.322	0	0	1	1	1	1	2	2	2
5.4	2.324	2.326	2.328	2.330	2.332	2.335	2.337	2.339	2.341	2.343	0	0	1	1	1	1	2	2	2

Note: Columns 1–9 on the right are the **Mean Differences**.

SQUARE ROOTS

	0	1	2	3	4	5	6	7	8	9	Mean Differences 1 2 3	4 5 6	7 8 9
5·5	2·345	2·347	2·349	2·352	2·354	2·356	2·358	2·360	2·362	2·364	0 0 1	1 1 1	1 2 2
5·6	2·366	2·369	2·371	2·373	2·375	2·377	2·379	2·381	2·383	2·385	0 0 1	1 1 1	1 2 2
5·7	2·387	2·390	2·392	2·394	2·396	2·398	2·400	2·402	2·404	2·406	0 0 1	1 1 1	1 2 2
5·8	2·408	2·410	2·412	2·415	2·417	2·419	2·421	2·423	2·425	2·427	0 0 1	1 1 1	1 2 2
5·9	2·429	2·431	2·433	2·435	2·437	2·439	2·441	2·443	2·445	2·447	0 0 1	1 1 1	1 2 2
6·0	2·449	2·452	2·454	2·456	2·458	2·460	2·462	2·464	2·466	2·468	0 0 1	1 1 1	1 2 2
6·1	2·470	2·472	2·474	2·476	2·478	2·480	2·482	2·484	2·486	2·488	0 0 1	1 1 1	1 2 2
6·2	2·490	2·492	2·494	2·496	2·498	2·500	2·502	2·504	2·506	2·508	0 0 1	1 1 1	1 2 2
6·3	2·510	2·512	2·514	2·516	2·518	2·520	2·522	2·524	2·526	2·528	0 0 1	1 1 1	1 2 2
6·4	2·530	2·532	2·534	2·536	2·538	2·540	2·542	2·544	2·546	2·548	0 0 1	1 1 1	1 2 2
6·5	2·550	2·551	2·553	2·555	2·557	2·559	2·561	2·563	2·565	2·567	0 0 1	1 1 1	1 2 2
6·6	2·569	2·571	2·573	2·575	2·577	2·579	2·581	2·583	2·585	2·587	0 0 1	1 1 1	1 2 2
6·7	2·588	2·590	2·592	2·594	2·596	2·598	2·600	2·602	2·604	2·606	0 0 1	1 1 1	1 2 2
6·8	2·608	2·610	2·612	2·613	2·615	2·617	2·619	2·621	2·623	2·625	0 0 1	1 1 1	1 2 2
6·9	2·627	2·629	2·631	2·632	2·634	2·636	2·638	2·640	2·642	2·644	0 0 1	1 1 1	1 2 2
7·0	2·646	2·648	2·650	2·651	2·653	2·655	2·657	2·659	2·661	2·663	0 0 1	1 1 1	1 2 2
7·1	2·665	2·666	2·668	2·670	2·672	2·674	2·676	2·678	2·680	2·681	0 0 1	1 1 1	1 1 2
7·2	2·683	2·685	2·687	2·698	2·691	2·693	2·694	2·696	2·698	2·700	0 0 1	1 1 1	1 1 2
7·3	2·702	2·704	2·706	2·707	2·709	2·711	2·713	2·715	2·717	2·718	0 0 1	1 1 1	1 1 2
7·4	2·720	2·722	2·724	2·726	2·728	2·729	2·731	2·733	2·735	2·737	0 0 1	1 1 1	1 1 2
7·5	2·739	2·740	2·742	2·744	2·746	2·748	2·750	2·751	2·753	2·755	0 0 1	1 1 1	1 1 2
7·6	2·757	2·759	2·760	2·762	2·764	2·766	2·768	2·769	2·771	2·773	0 0 1	1 1 1	1 1 2
7·7	2·775	2·777	2·778	2·780	2·782	2·784	2·786	2·787	2·789	2·791	0 0 1	1 1 1	1 1 2
7·8	2·793	2·795	2·796	2·798	2·800	2·802	2·804	2·805	2·807	2·809	0 0 1	1 1 1	1 1 2
7·9	2·811	2·812	2·814	2·816	2·818	2·820	2·821	2·823	2·825	2·827	0 0 1	1 1 1	1 1 2
8·0	2·828	2·830	2·832	2·834	2·835	2·837	2·839	2·841	2·843	2·844	0 0 1	1 1 1	1 1 2
8·1	2·846	2·848	2·850	2·851	2·853	2·855	2·857	2·858	2·860	2·862	0 0 1	1 1 1	1 1 2
8·2	2·864	2·865	2·867	2·869	2·871	2·872	2·874	2·876	2·877	2·879	0 0 1	1 1 1	1 1 2
8·3	2·881	2·883	2·884	2·886	2·888	2·890	2·891	2·893	2·895	2·897	0 0 1	1 1 1	1 1 2
8·4	2·898	2·900	2·902	2·903	2·905	2·907	2·909	2·910	2·912	2·914	0 0 1	1 1 1	1 1 2
8·5	2·915	2·917	2·919	2·921	2·922	2·924	2·926	2·927	2·929	2·931	0 0 1	1 1 1	1 1 2
8·6	2·933	2·934	2·936	2·938	2·939	2·941	2·943	2·944	2·946	2·948	0 0 1	1 1 1	1 1 2
8·7	2·950	2·951	2·953	2·955	2·956	2·958	2·960	2·961	2·963	2·965	0 0 1	1 1 1	1 1 2
8·8	2·966	2·968	2·970	2·972	2·973	2·975	2·977	2·978	2·980	2·982	0 0 1	1 1 1	1 1 2
8·9	2·983	2·985	2·987	2·988	2·990	2·992	2·993	2·995	2·997	2·998	0 0 1	1 1 1	1 1 2
9·0	3·000	3·002	3·003	3·005	3·007	3·008	3·010	3·012	3·013	3·015	0 0 0	1 1 1	1 1 1
9·1	3·017	3·018	3·020	3·022	3·023	3·025	3·027	3·028	3·030	3·032	0 0 0	1 1 1	1 1 1
9·2	3·033	3·035	3·036	3·038	3·040	3·041	3·043	0·045	3·046	3·048	0 0 0	1 1 1	1 1 1
9·3	3·050	3·051	3·053	3·055	3·056	3·058	3·059	3·061	3·063	3·064	0 0 0	1 1 1	1 1 1
9·4	3·066	3·068	3·069	3·071	3·072	3·074	3·076	3·077	3·079	3·081	0 0 0	1 1 1	1 1 1
9·5	3·082	3·084	3·085	3·087	3·089	3·090	3·092	3·094	3·095	3·097	0 0 0	1 1 1	1 1 1
9·6	3·098	3·100	3·102	3·103	3·105	3·106	3·108	3·110	3·111	3·113	0 0 0	1 1 1	1 1 1
9·7	3·114	3·116	3·118	3·119	3·121	3·122	3·124	3·126	3·127	3·129	0 0 0	1 1 1	1 1 1
9·8	3·130	3·132	3·134	3·135	3·137	3·138	3·140	3·142	3·145	3·145	0 0 0	1 1 1	1 1 1
9·9	3·146	3·148	3·150	3·151	3·153	3·154	3·156	3·158	3·159	3·161	0 0 0	1 1 1	1 1 1

SQUARE ROOTS

	0	1	2	3	4	5	6	7	8	9	Mean Differences		
											1 2 3	4 5 6	7 8 9
10	3·162	3·178	3·194	3·209	3·225	3·240	3·256	3·271	3·286	3·302	2 3 5	6 8 9	11 12 14
11	3·317	3·332	3·347	3·362	3·376	3·391	3·406	3·421	3·435	3·450	1 3 4	6 7 9	10 12 13
12	3·464	3·479	3·493	3·507	3·521	3·536	3·550	3·564	3·578	3·592	1 3 4	6 7 8	10 11 13
13	3·606	3·619	3·633	3·647	3·661	3·674	3·688	3·701	3·715	3·728	1 3 4	5 7 8	10 11 12
14	3·742	3·755	3·768	3·782	3·795	3·808	3·821	3·834	3·847	3·860	1 3 4	5 7 8	9 11 12
15	3·873	3·886	3·899	3·912	3·924	3·937	3·950	3·962	3·975	3·987	1 3 4	5 6 8	9 10 11
16	4·000	4·012	4·025	4·037	4·050	4·062	4·074	4·087	4·099	4·411	1 2 4	5 6 7	9 10 11
17	4·123	4·135	4·147	4·159	4·171	4·183	4·195	4·207	4·219	4·231	1 2 4	5 6 7	8 10 11
18	4·243	4·254	4·266	4·278	4·290	4·301	4·313	4·324	4·336	4·347	1 2 3	5 6 7	8 9 10
19	4·359	4·370	4·382	4·393	4·405	4·416	4·427	4·438	4·450	4·461	1 2 3	5 6 7	8 9 10
20	4·472	4·483	4·494	4·506	4·517	4·528	4·539	4·550	4·561	4·572	1 2 3	4 6 7	8 9 10
21	4·583	4·593	4·604	4·615	4·626	4·637	4·648	4·658	4·669	4·680	1 2 3	4 5 6	8 9 10
22	4·690	4·701	4·712	4·722	4·733	4·743	4·754	4·764	4·775	4·785	1 2 3	4 5 6	7 8 9
23	4·796	4·806	4·817	4·827	4·837	4·848	4·858	4·868	4·879	4·889	1 2 3	4 5 6	7 8 9
24	4·899	4·909	4·919	4·930	4·940	4·950	4·960	4·970	4·980	4·990	1 2 3	4 5 6	7 8 9
25	5·000	5·010	5·020	5·030	5·040	5·050	5·060	5·070	5·079	5·089	1 2 3	4 5 6	7 8 9
26	5·099	5·109	5·119	5·128	5·138	5·148	5·158	5·167	5·177	5·187	1 2 3	4 5 6	7 8 9
27	5·196	5·206	5·215	5·225	5·235	5·244	5·254	5·263	5·273	5·282	1 2 3	4 5 6	7 8 9
28	5·292	5·301	5·310	5·320	5·329	5·339	5·348	5·357	5·367	5·376	1 2 3	4 5 6	7 7 8
29	5·385	5·394	5·404	5·413	5·422	5·431	5·441	5·450	5·459	5·468	1 2 3	4 5 5	6 7 8
30	5·477	5·486	5·495	5·505	5·514	5·523	5·532	5·541	5·550	5·559	1 2 3	4 4 5	6 7 8
31	5·568	5·577	5·586	5·595	5·604	5·612	5·621	5·630	5·639	5·648	1 2 3	3 4 5	6 7 8
32	5·657	5·666	5·675	5·683	5·692	5·701	5·710	5·718	5·727	5·736	1 2 3	3 4 5	6 7 8
33	5·745	5·753	5·771	5·779	5·779	5·788	5·797	5·805	5·814	5·822	1 2 3	3 4 5	6 7 8
34	5·831	5·840	5·848	5·857	5·865	5·874	5·882	5·891	5·899	5·908	1 2 3	3 4 5	6 7 8
35	5·916	5·925	5·933	5·941	5·950	5·958	5·967	5·975	5·983	5·992	1 2 2	3 4 5	6 7 8
36	6·000	6·008	6·017	6·025	6·033	6·042	6·050	6·058	6·066	6·075	1 2 2	3 4 5	6 7 7
37	6·083	6·091	6·099	6·107	6·116	6·124	6·132	6·140	6·148	6·156	1 2 2	3 4 5	6 7 7
38	6·164	6·173	6·181	6·189	6·197	6·205	6·213	6·221	6·229	6·237	1 2 2	3 4 5	6 6 7
39	6·245	6·253	6·261	6·269	6·277	6·285	6·293	6·301	6·309	6·317	1 2 2	3 4 5	6 6 7
40	6·625	6·332	6·340	6·348	6·356	6·364	6·372	6·380	6·387	6·395	1 2 3	3 4 5	6 6 7
41	6·403	6·411	6·419	6·427	6·434	6·442	6·450	6·458	6·465	6·473	1 2 3	3 4 5	5 6 7
42	6·481	6·488	6·496	6·504	6·512	6·519	6·527	6·535	6·542	6·550	1 2 2	3 4 5	5 6 7
43	6·557	6·565	6·573	6·580	6·588	6·595	6·603	6·611	6·618	6·626	1 2 2	3 4 5	5 6 7
44	6·633	6·641	6·648	6·656	6·663	6·671	6·678	6·686	6·693	6·701	1 2 2	3 4 5	5 6 7
45	6·708	6·716	6·723	6·731	6·738	6·745	6·753	6·760	6·768	6·775	1 1 2	3 4 4	5 6 7
46	6·782	6·790	6·797	6·804	6·812	6·819	6·826	6·834	6·841	6·848	1 1 2	3 4 4	5 6 7
47	6·856	6·863	6·870	6·877	6·885	6·892	6·899	6·907	6·914	6·921	1 1 2	3 4 4	5 6 7
48	6·928	6·935	6·943	6·950	6·957	6·964	6·971	6·979	6·986	6·993	1 1 2	3 4 4	5 6 6
49	7·000	7·007	7·014	7·021	7·029	7·036	7·043	7·050	7·057	7·064	1 1 2	3 4 4	5 6 6
50	7·071	7·078	7·085	7·092	7·099	7·106	7·113	7·120	7·127	7·134	1 1 2	3 4 4	5 6 6
51	7·141	7·148	7·155	7·162	7·176	7·176	7·183	7·190	7·197	7·204	1 1 2	3 4 4	5 6 6
52	7·211	7·218	7·225	7·232	7·239	7·246	7·253	7·259	7·266	7·273	1 1 2	3 3 4	5 6 6
53	7·280	7·287	7·294	7·301	7·308	7·314	7·321	7·328	7·335	7·342	1 1 2	3 3 4	5 5 6
54	7·348	7·355	7·362	7·369	7·376	7·382	7·396	7·396	7·403	7·409	1 1 2	3 3 4	5 5 6

SQUARE ROOTS

	0	1	2	3	4	5	6	7	8	9	Mean Differences								
											1	2	3	4	5	6	7	8	9
55	7·410	7·423	7·430	7·436	7·443	7·450	7·457	7·463	7·470	7·477	1	1	2	3	3	4	5	5	6
56	7·483	7·490	7·497	7·503	7·510	7·517	7·523	7·530	7·537	7·543	1	1	2	3	3	4	5	5	6
57	7·550	7·556	7·563	7·570	7·576	7·583	7·589	7·596	7·603	7·609	1	1	2	3	3	4	5	5	6
58	7·616	7·622	7·629	7·635	7·642	7·649	7·655	7·662	7·668	7·675	1	1	2	3	3	4	5	5	6
59	7·681	7·688	7·694	7·701	7·707	7·714	7·720	7·727	7·733	7·740	1	1	2	3	3	4	4	5	6
60	7·746	7·752	7·759	7·765	7·772	7·778	7·785	7·791	7·797	7·804	1	1	2	3	3	4	4	5	6
61	7·810	7·817	7·823	7·829	7·836	7·842	7·849	7·855	7·861	7·868	1	1	2	3	3	4	4	5	6
62	7·874	7·880	7·887	7·893	7·899	7·906	7·912	7·918	7·925	7·931	1	1	2	3	3	4	4	5	6
63	7·937	7·944	7·950	7·956	7·962	7·969	7·975	7·981	7·987	7·994	1	1	2	3	3	4	4	5	6
64	8·000	8·006	8·012	8·019	8·025	8·031	8·037	8·044	8·050	8·056	1	1	2	2	3	4	4	5	6
65	8·062	8·068	8·075	8·081	8·087	8·093	8·099	8·106	8·112	8·118	1	1	2	2	3	4	4	5	6
66	8·124	8·130	8·136	8·142	8·149	8·155	8·161	8·167	8·173	8·179	1	1	2	2	3	4	4	5	5
67	8·185	8·191	8·198	8·204	8·210	8·216	8·222	8·228	8·234	8·240	1	1	2	2	3	4	4	5	5
68	8·246	8·252	8·258	8·264	8·270	8·276	8·283	8·289	8·295	8·301	1	1	2	2	3	4	4	5	5
69	8·307	8·313	8·319	8·325	8·331	8·337	8·343	8·349	8·355	8·361	1	1	2	2	3	4	4	5	5
70	8·367	8·373	8·379	8·385	8·390	8·396	8·402	8·408	8·414	8·420	1	1	2	2	3	4	4	5	5
71	8·426	8·432	8·438	8·444	8·450	8·456	8·462	8·468	8·473	8·479	1	1	2	2	3	4	4	5	5
72	8·485	8·491	8·497	8·503	8·509	8·515	8·521	8·526	8·532	8·538	1	1	2	2	3	3	4	5	5
73	8·544	8·550	8·556	8·562	8·576	8·573	8·579	8·585	8·591	8·597	1	1	2	2	3	3	4	5	5
74	8·602	8·608	8·614	8·620	8·626	8·631	8·637	8·643	8·649	8·654	1	1	2	2	3	3	4	5	5
75	8·660	8·666	8·672	8·678	8·683	8·689	8·695	8·701	8·706	8·712	1	1	2	2	3	3	4	5	5
76	8·718	8·724	8·729	8·735	8·741	8·746	8·752	8·758	8·764	8·769	1	1	2	2	3	3	4	5	5
77	8·775	8·781	8·786	8·792	8·798	8·803	8·809	8·815	8·820	8·826	1	1	2	2	3	3	4	4	5
78	8·832	8·837	8·843	8·849	8·854	8·860	8·866	8·871	8·877	8·883	1	1	2	2	3	3	4	4	5
79	8·888	8·894	8·899	8·905	8·911	8·916	8·922	8·927	8·933	8·939	1	1	2	2	3	3	4	4	5
80	8·944	8·950	8·955	8·961	8·967	8·972	8·978	8·983	8·989	8·994	1	1	2	2	3	3	4	4	5
81	9·000	9·006	9·001	9·017	9·022	9·028	9·033	9·039	9·044	9·050	1	1	2	2	3	3	4	4	5
82	9·055	9·061	9·066	9·072	9·077	9·083	9·088	9·094	9·099	9·105	1	1	2	2	3	3	4	4	5
83	9·110	9·116	9·121	9·127	9·132	9·138	9·143	9·149	9·154	9·160	1	1	2	2	3	3	4	4	5
84	9·165	9·171	9·176	9·182	9·187	9·192	9·198	9·203	9·209	9·214	1	1	2	2	3	3	4	4	5
85	9·220	9·225	9·230	9·236	9·241	9·247	9·252	9·257	9·263	9·268	1	1	2	2	3	3	4	4	5
86	9·274	9·279	9·284	9·290	9·295	9·301	9·306	9·311	9·317	9·322	1	1	2	2	3	3	4	4	5
87	9·327	9·333	9·338	9·343	9·349	9·354	9·359	9·365	9·370	9·375	1	1	2	2	3	3	4	4	5
88	9·381	9·386	9·391	9·397	9·402	9·407	9·413	9·418	9·423	9·429	1	1	2	2	3	3	4	4	5
89	9·434	9·439	9·445	9·450	9·455	9·460	9·466	9·471	9·476	9·482	1	1	2	2	3	3	4	4	5
90	9·487	9·492	9·497	9·503	9·508	9·513	9·518	9·524	9·529	9·534	1	1	2	2	3	3	4	4	5
91	9·539	9·545	9·550	9·555	9·560	9·566	9·571	9·576	9·581	9·586	1	1	2	2	3	3	4	4	5
92	9·592	9·597	9·602	9·607	9·612	9·618	9·623	9·628	9·633	9·638	1	1	2	2	3	3	4	4	5
93	9·644	9·649	9·654	9·659	9·664	9·670	9·675	9·680	9·685	9·690	1	1	2	2	3	3	4	4	5
94	9·695	9·701	9·706	9·711	9·716	9·721	9·726	9·731	9·737	9·742	1	1	2	2	3	3	4	4	5
95	9·747	9·752	9·757	9·762	9·767	9·772	9·778	9·783	9·788	9·793	1	1	2	2	3	3	4	4	5
96	9·798	9·803	9·808	9·813	9·818	9·823	9·829	9·834	9·839	9·844	1	1	2	2	3	3	4	4	5
97	9·849	9·854	9·859	9·864	9·869	9·874	9·879	9·884	9·889	9·894	1	1	1	2	3	3	4	4	5
98	9·899	9·905	9·910	9·915	9·920	9·925	9·930	9·935	9·940	9·945	1	1	1	2	2	3	3	4	4
99	9·950	9·995	9·960	9·965	9·970	9·975	9·980	9·985	9·990	9·995	0	1	1	2	2	3	3	4	4

Reproduced here from Frank Castle's *Four-Figure Mathematical Tables,*
Macmillan

3.3 Answers to exercises

Exercise 1.1 Page 10

1.	0·035	2.	22 000	3.	0·482
4.	0·047	5.	0·047	6.	50 000
7.	7	8.	20	9.	0·04
10.	3·6	11.	570	12.	3·42
13.	0·8725	14.	96·25	15.	0·012 35
16.	512	17.	326	18.	6100
19.	3·628	20.	7·9562	21.	64 000
22.	0·092	23.	325 000	24.	0·036
25.	54 300 000	26.	0·000 000 08	27.	0·000 006
28.	0·000 05	29.	0·0008	30.	0·039
31.	4·8	32.	87 400	33.	3218
34.	8258	35.	6 452 000	36.	0·000 768
37.	9·13	38.	0·000 000 008 4	39.	6 452 000
40.	8 294 600				

Exercise 1.2 Page 16

1.	7·28 m^2	2.	440 mm	3.	74·83 mm
4.	308 mm	5.	14·4 cm^2	6.	40 mm
7.	374·4 mm^2	8.	179 cm^2	9.	3326·4 cm^3
10.	12·54 cm^3	11.	(a) 10:1 (b) 12·25:1	12.	38 808 mm^3,
13.	70 mm	14.	356·4 ℓ		310·464 g
15.	20 ℓ	16.	3720 mm^2	17.	45°
18.	6·1 mm	19.	31°	20.	(a) 16°, 56°, 5
					(c) 252° Inlet

Exercise 1.3 Page 22

1. $\sqrt{(4A/\pi)}$ 2. $\sqrt[3]{\dfrac{3A}{4\pi}}$ 3. $\sqrt{(c^2 - b^2}$

4. $\dfrac{2A}{h} - 4$ 5. $\sqrt{\dfrac{A}{\pi} + r^2}$ 6. $V_c (R - 1)$

7. $\dfrac{V_s}{R-1}$ 8. $\dfrac{V_s}{V_c} + 1$ OR $\dfrac{V_s + V_c}{V_c}$ 9. $\dfrac{P}{2\pi n}$

10. $\dfrac{T}{S\mu r}$ 11. N m   12. m^3

13. g 14. N/mm^2

Exercise 1.4 Page 26

1. (a) 55 N m (b) 2500 rev/min   2. 3700–4700 rev

3. (a) CO_2 8 per cent, CO 9·5 per cent   4. (a) 21·5° (b) 11
   O_2 0 per cent; (b) 15:1

5. (a) 46 °C   6. (a) 8·2°
   (b) 66 °C   (b) 16°

7. (a) 81 km   8. (a) 3·25°
   (b) 140 km   (b) 194 mm
   (c) 48 km/h
   (d) 113 km/h

3 Tables and Answers

Exercise 1.5 Page 30

1.	0·1 mm	2.	0·02 mm	3.	0·02 mm
4.	4·5 mm	5.	52·7 mm	6.	125·7 mm
7.	1·68 mm	8.	43·24 mm	9.	122·46 mm
10.	5·48 mm	11.	93·68 mm	12.	151·76 mm

Exercise 2.1 Page 41

1.	2 kJ/K	2.	8 kg	3.	384 kJ
4.	840 kJ	5.	260 °C	6.	90°C
7.	3360 kJ	8.	1020 kJ	9.	38 °C
10.	21 °C				

Exercise 2.2 Page 43

1.	2 mm	2.	120·72 mm	3.	Aluminium alloy
4.	515 °C	5.	Slightly above 182°C	6.	80·064 mm
7.	80·4 mm	8.	5050 mm^2	9.	6656 cm^3
10.	0·192 ℓ				

Exercise 2.3 Page 48

1.	310 K	2.	310 C(abs)	3.	75 cm^3
4.	687 °C	5.	7 bar OR 700 kPa	6.	9 bar OR 900 kPa
7.	8 bar OR 800 kPa	8.	1000 K	9.	4·7 bar OR 470 kPa
10.	291 °C				

Exercise 2.4 Page 51

1. (a) 168 kJ (b) 42 kJ (c) 1680 kJ
   (d) 6780 kJ (e) 705·6 kJ (f) 13 610·4 kJ

Exercise 2.26 Page 98

1.	0·6	2.	5·5 kN
3.	4·32 Nm	4.	0·35
5.	140 Nm	6.	115·2 Nm
7.	70·4 kW	8.	900 N
9.	175 N	10.	3·2 Nm
11.	100 N	12.	12 kN
13.	2 m from front wheels	14.	0·5 m from front wheel
15.	180 mm from small end	16.	172 rev/min; 1·26 kNm
17.	60	18.	(a) 350 rev/min, (b) 787·5 N m
19.	75 per cent	20.	(a) 40 N, (b) 21·12 m/s, (c) 2 Nm at 67·2 rev/s
21.	1400 rev/min	22.	3:1
23.	2700 rev/min	24.	(a) 3·5:1 (b) 1·4:1 (c) 2·5:1
25.	9 kN	26.	A = 750 N B = 875 N
27.	(a) 1·5 mm (b) 12 mm	28.	337·5 N
29.	12 m/s	30.	27 m/s
31.	2·6 m/s^2	32.	70 per cent
33.	4 s	34.	48 m
35.	65 per cent	36.	(a) 540 kJ (b) 60 m (c) 75 per cent

3.4 Answers to multiple choice questions

Section 1 (1.6) Page 31

1.	(a)	2.	(d)	3.	(a)	4.	(d)	5.	(a
6.	(b)	7.	(d)	8.	(c)	9.	(a)	10.	(d
11.	(a)	12.	(a)	13.	(d)	14.	(c)	15.	(b
16.	(b)	17.	(c)	18.	(a)	19.	(b)	20.	(c
21.	(b)	22.	(c)	23.	(b)	24.	(b)	25.	(d
26.	(d)	27.	(d)	28.	(c)	29.	(d)	30.	(d
31.	(a)	32.	(b)	33.	(c)	34.	(c)	35.	(a
36.	(a)	37.	(b)	38.	(d)	39.	(c)	40.	(c
41.	(c)	42.	(d)	43.	(b)	44.	(c)	45.	(c
46.	(c)	47.	(d)	48.	(c)	49.	(d)	50.	(c
51.	(b)	52.	(d)	53.	(a)	54.	(b)	55.	(a
56.	(c)	57.	(b)	58.	(a)	59.	(d)	60.	(b
61.	(d)								

Section 2(a) (2.5) Page 51

1.	(a)	2.	(d)	3.	(c)	4.	(d)	5.	(c
6.	(d)	7.	(c)	8.	(b)	9.	(b)	10.	(a
11.	(d)	12.	(c)	13.	(a)	14.	(a)	15.	(
16.	(b)	17.	(d)	18.	(c)	19.	(c)	20.	(a
21.	(c)	22.	(c)	23.	(c)	24.	(b)	25.	(c
26.	(b)	27.	(b)	28.	(b)	29.	(b)	30.	(b
31.	(d)	32.	(a)	33.	(c)	34.	(b)	35.	(c
36.	(b)	37.	(c)	38.	(c)	39.	(a)	40.	(c
41.	(d)	42.	(c)	43.	(a)	44.	(b)	45.	(
46.	(d)	47.	(c)	48.	(b)				

Section 2(b) (2.14) Page 67

1.	(a)	2.	(c)	3.	(b)	4.	(c)	5.	(c
6.	(b)	7.	(d)	8.	(c)	9.	(c)	10.	(a
11.	(a)	12.	(d)	13.	(a)	14.	(a)	15.	(d
16.	(d)	17.	(a)	18.	(c)	19.	(b)	20.	(c
21.	(c)	22.	(c)	23.	(a)	24.	(c)	25.	(c
26.	(d)	27.	(a)	28.	(a)	29.	(b)	30.	(c
31.	(d)	32.	(b)	33.	(c)	34.	(b)	35.	(c
36.	(b)	37.	(b)	38.	(b)	39.	(c)	40.	(b

Section 2(c) (2.26) Page 101

1.	(c)	2.	(d)	3.	(d)	4.	(b)	5.	(b
6.	(b)	7.	(d)	8.	(c)	9.	(d)	10.	(c
11.	(b)	12.	(c)	13.	(a)	14.	(d)	15.	(c
16.	(d)	17.	(a)	18.	(b)	19.	(b)	20.	(a
21.	(c)	22.	(a)	23.	(d)	24.	(d)	25.	(c
26.	(a)	27.	(b)	28.	(b)	29.	(c)	30.	(c)
31.	(b)	32.	(c)	33.	(a)	34.	(c)	35.	(a)
36.	(d)	37.	(a)	38.	(c)	39.	(b)	40.	(c)
41.	(b)	42.	(c)	43.	(a)	44.	(c)	45.	(d)
46.	(a)	47.	(a)	48.	(b)	49.	(d)	50.	(d)
51.	(c)	52.	(c)	53.	(c)	54.	(a)	55.	(a)
56.	(a)	57.	(a)	58.	(c)	59.	(c)	60.	(d
61.	(a)	62.	(b)	63.	(d)	64.	(d)	65.	(b
66.	(b)	67.	(d)	68.	(c)	69.	(a)	70.	(

Section 2(d) (2.40) Page 137

1. (d)	2. (b)	3. (b)	4. (b)	5. (a)
6. (d)	7. (c)	8. (a)	9. (c)	10. (b)
11. (c)	12. (b)	13. (d)	14. (d)	15. (d)
16. (a)	17. (b)	18. (c)	19. (a)	20. (d)
21. (a)	22. (b)	23. (a)	24. (c)	25. (b)
26. (d)	27. (b)	28. (c)	29. (a)	30. (d)
31. (c)	32. (c)	33. (a)	34. (c)	35. (a)
36. (a)	37. (d)	38. (d)	39. (a)	40. (a)
41. (c)	42. (a)	43. (b)	44. (a)	45. (d)
46. (a)	47. (c)	48. (d)	49. (d)	50. (d)
51. (b)	52. (a)	53. (b)	54. (d)	55. (c)
56. (c)	57. (d)	58. (a)	59. (b)	60. (d)
61. (a)	62. (c)	63. (b)	64. (a)	65. (a)
66. (d)	67. (a)	68. (c)	69. (c)	70. (a)
71. (b)	72. (c)			

Section 2(e) (2.52) Page 150

1. (d)	2. (d)	3. (b)	4. (c)	5. (b)
6. (a)	7. (a)	8. (d)	9. (a)	10. (b)
11. (c)	12. (c)	13. (a)	14. (d)	15. (b)
16. (d)	17. (d)	18. (b)	19. (d)	20. (c)
21. (c)	22. (d)	23. (d)	24. (d)	25. (a)

INDEX